I0121706

Biotechnology in Japan

In the early 1980s, biotechnology caused worldwide excitement as a high technology with almost unlimited potential in science, medicine, and industry. It not only allowed the manufacture of traditional products more quickly and inexpensively, but also offered the possibility of synthesizing valuable materials anew. All this made it very attractive to Japanese policy-makers, who 'targeted' it as a high priority area for economic growth.

Originally published in 1989, *Biotechnology in Japan* is the first published in English to analyse the Japanese effort to promote the new biotechnology industries. The author examines the strategies used for developing biotechnology in Japan and looks at the active role of government in a field in which the Japanese rapidly became the world leaders. Focusing on the making and implementation of biotechnology policy, he considers the relationship between the public and the private sector, and makes use of different political constructs to analyse Japan's complex and unique balance between competitive market forces and collective interest.

Biotechnology in Japan

Malcolm V. Brock

Routledge
Taylor & Francis Group

First published in 1989
by Routledge

This edition first published in 2024 by Routledge
4 Park Square, Milton Park, Abingdon, Oxon, OX14 4RN

and by Routledge
605 Third Avenue, New York, NY 10017

Routledge is an imprint of the Taylor & Francis Group, an informa business

Publisher's Note
The publisher has gone to great lengths to ensure the quality of this reprint but points out that some imperfections in the original copies may be apparent.

Disclaimer
The publisher has made every effort to trace copyright holders and welcomes correspondence from those they have been unable to contact.

A Library of Congress record exists under LCCN: 89005966

ISBN: 978-1-032-89788-2 (hbk)
ISBN: 978-1-003-54460-9 (ebk)
ISBN: 978-1-032-89791-2 (pbk)

Book DOI 10.4324/9781003544609

Biotechnology in Japan

Malcolm V. Brock

R
Routledge
London and New York

First published 1989 by Routledge
11 New Fetter Lane, London EC4P 4EE
29 West 35th Street, New York, NY 10001

Typeset by J&L Composition Ltd, Filey, North Yorkshire
Printed and bound in Great Britain by
Biddles Ltd, Guildford and King's Lynn

British Library Cataloguing in Publication Data

Brock, Malcolm V.
   Biotechnology in Japan. —— (Nissan
   Institute/Routledge Japanese studies
   series)
   1. Japan. Biotechnology industries
   I. Title   II. Series
   338.4'76606'0952
   ISBN 0-415-03495-7

Library of Congress Cataloging-in-Publication Data

Brock, Malcolm Vernon, 1964–
   Biotechnology in Japan / Malcolm Vernon Brock.
   p.   cm. — (The Nissan Institute/Routledge Japanese studies
   series)
   Bibliography: p.
   ISBN 0-415-03495-7
   1. Biotechnology—Government policy—Japan. 2. Biotechnology—
   Economic aspects—Japan. 3. Biotechnology industries—Government
   policy—Japan. I. Title. II. Series.
   TP248.195.J3B76 1989
   338.4'76208'0952—dc19
                                 89-5966
                                    CIP

Dedication

To Mom and Dad.
Once is not enough.

Contents

General Editor's Preface

Almost imperceptibly, during the 1980s, Japan has become 'hot news'. The successes of the Japanese economy and the resourcefulness of her people have long been appreciated abroad. What is new is an awareness of her increasing impact on the outside world. This tends to produce painful adjustment and uncomfortable reactions. It also often leads to stereotypes and arguments based on outdated or ill-informed ideas.

The Nissan Institute/Routledge Japanese Studies Series seeks to foster an informed and balanced – but not uncritical – understanding of Japan. One aim of the series is to show the depth and variety of Japanese institutions, practices and ideas. Another is, by using comparison, to see what lessons, positive and negative, can be drawn for other countries. There are many aspects of Japan which are little known outside that country but which deserve to be better understood.

High technology research and development is now a startlingly crucial component of Japan's economic dynamism. Is is a long time since the Japanese could be regarded as a nation of industrial imitators, and innovation is granted the highest priority. Malcolm Brock's study focuses on biotechnology, an area where Japan can draw on traditional expertise in techniques such as those used in making *sake*, and where the latest in scientific know-how is being applied towards new and productive advances. Mr Brock focuses on interactions between vigorously competing forms, supervision of developmental policies by the state, rivalry and co-operation between government ministries, and the contribution of academics. He paints a picture of complex and dynamic interaction.

J. J. A. Stockwin
Director, Nissan Institute of Japanese Studies,
University of Oxford

Acknowledgements

I would like to thank Professor J. A. A. Stockwin for his patience and understanding in supervising me during my two years at Oxford. He has been not only a constant inspiration with memos, valuable articles and teaching, but also has given me the opportunity to meet with eminent scholars in Japanese Studies, and to participate in seminars both in and outside of Oxford. He has supported me every step of the way, and for this I am very grateful.

I am also deeply indebted to Dr Kent Calder at the Woodrow Wilson School, Princeton University. It was he who initially gave me the confidence and encouragement to apply to Oxford and tackle this work.

Special thanks also to those in Japan who made my field trip productive and enlightening. Special mention to Morishita Noboru who sat and debated issues for many hours, and to Professor Saitō Hyūga who served as my mentor while I was in Tokyo. Last, but not least, I thank my father for burning the midnight oil while helping me to edit in the final stages.

Thanks are due also to the following who granted me interviews:

Aiba Yasuhide. Special R&D Policy Office of the Agriculture, Forestries and Fisheries Council. Planning Strategist; 24 September 1986.

Fujimura Robert. First Secretary (Biological Sciences) American Embassy; 1 September and 21 October 1986.

Ikegaya Sōichi. Ministry of Health and Welfare, Deputy Director of the Life Science Office; 17 September and 30 October 1986.

Kobayashi Shinichi. Deputy Director of the Ministry of Agriculture, Forestries and Fisheries' Communications Office; 11 September 1986.

Acknowledgements

Kurata Kenji. MITI's Bioindustry Office Officer; September 1985.

Miyata Mitsuru. *Nikkei Baioteku* Editor; 31 October 1986.

Morishita Noboru. journalist *Nihon Kōgyō Shimbun*; 28 August and 20 October 1986.

Motoshima Naoki. Deputy Director of Technology Research and Information Division, MITI's Agency of Industrial Science and Technology; 21 October 1986.

Nakata Yusaku. MITI's Agency of Industrial Science and Technology, Office for the Promotion of Basic R&D; 22 September 1986.

Ōita Takahisa. Professor of Agriculture at Tokyo University; 21 October 1986.

Ōyama Chō. Director of MOE's Research Subsidy Division; 11 September 1986.

Saitō Hyūga. Professor of Microbiology at Tokyo University, Director of the Applied Microbiology Institute; 14 August and 3 October 1986.

Hirotoshi, Samejima. Executive Vice President Kyowa Medex Co., Ltd.; 24 September 1986.

Sasaki Shuichi. Director for Future Industrial Technology at MITI's Agency of Industrial Science and Technology; 26 September 1986.

Satō Masuke. Science and Technology Agency's Life Science Officer; 9 September 1986.

Shimizu Makoto. Managing Director of the Bioindustry Development Center (BIDEC); 16 September 1986.

Tanaka Masami. Director of the Technology Development Division in the Science and Technology Agency (former Director of MITI's Bioindustry Office); 12 September 1986.

List of Abbreviations

AB	Advisory Body
AIST	Agency of Industrial Science and Technology
BAC	Bioindustry Advisory Council
BIDEC	Bioindustry Development Centre
BRAIN	Biotechnology Research Advancement Institution
CST	Council on Science and Technology
DNA	Deoxyribonucleic Acid
ERATO	Exploratory Research and Advanced Technology Organisation
GHQ	General Headquarters
JECC	Japan Electric Computer Corporation
JSC	Japan Science Council
LDP	Liberal Democratic Party
MAFF	Ministry of Agriculture, Forestries and Fisheries
MHW	Ministry of Health and Welfare
MITI	Ministry of International Trade and Industry
MOE	Ministry of Education
MOF	Ministry of Finance
MPT	Ministry of Posts and Telecommunications
NBF	New Biotechnology Firms
NEC	Nippon Electric Corporation
NGBT	Next Generation Basic Technologies
NIAR	National Institute of Agribiological Resources
NIH	National Institute of Health
NRIJU	National Research Institute for Joint Use by Universities
NTT	Nippon Telephone and Telegram
OTA	Office of Technology Assessment
PERI	Protein Engineering Research Institute
RA	Research Association
rDNA	Recombinant Deoxyribonucleic Acid
R&D	Research and Development
RIKEN	Physics and Chemistry Institute (*Ri*kagaku *Ken*kyūjo)

SCP	Single Cell Protein
SOPS	Standard Operating Procedures
STA	Science and Technology Agency
TPA	Tissue Plasminogen Activator
UNIDO	United Nations Independent Development Organisation
VLSI	Very Large-Scale Integrated Circuit
WHO	World Health Organisation

Introduction

In March 1980, the Japanese Ministry of International Trade and Industry (MITI) announced a long-term economic plan proposing high technology industries – more specifically microelectronics, new materials, and biotechnology – as the basis for economic growth in a future Japanese economy. Almost overnight, a multitude of government reports, mass-media accounts, and business portfolios pinpointed biotechnology specifically as the most important of the three, destined to produce valuable scientific advances in the coming decade. A mass education campaign promoted by the media heralded biotechnology as, 'the last technological revolution of the twentieth century', and forecasted that it would change all aspects of contemporary living, from clothes worn to food eaten, and ultimately the length of human life itself.

Experts and non-specialists alike asserted that once the full potential of biotechnology was realized, effects would be felt in agriculture, medicine, energy, the environment, and in the chemical, food and drink, and new-materials industries. This would mean new drugs to fight presently incurable diseases, recyclable sources of energy, more nutritious and less-expensive foods, fresh approaches to develop new and existing chemicals, and so on. The almost limitless potential of biotechnology for industrial applications and its promised savings in precious resources made it very attractive to Japanese policymakers. By mid 1982, it was clear that Japan, a nation notorious for its dependency on imported raw materials, energy, and food sources, had 'targeted' biotechnology as a means of both attaining economic growth and breaking a century-old reliance on the West.

In a broad sense, there have been two schools assessing postwar Japanese industrial policy. The first, often attributed to Chalmers Johnson and his book, *MITI and the Japanese Miracle*,

1

is often referred to as 'state-led capitalism' – a perspective attaching great importance to the state's role (mainly MITI) in promoting industrial change. The second, a favourite of *laissez-faire* economists, proposes the innovation of the private sector and responses to market changes as primary motive forces for rapid industrial growth. As so often happens in the social sciences, an adequate understanding of 'successful' Japanese industrial policy in iron and steel, shipbuilding, cars, as well as consumer electronics calls for an appreciation of some combination of these two theories.

This book attempts to clarify the grey area between the two schools by examining a case study of biotechnology. Indeed, three perspectives are presented as windows, if you will, through which a thorough analysis of Japanese biotechnology policy can be gained. These developmental frameworks attempt, first, to offer an hypothesis of the biotechnology policy process, second, to show the empirical evidence, and finally to construct a composite picture to fit the data collected.

The first of these hypothetical perspectives is *Nebukai* or the theory of policy practices being 'deeply rooted' in postwar Japanese political behaviour. *Nebukai* suggests that the relationships between the business community, the bureaucracy, and the ruling Liberal Democratic Party (LDP) have evolved over time into a normative, regularized pattern because of both a tradition of experiences in dealing with similar political and economic problems, and also an environment of stability due to the long, unbroken dominance of the LDP. The second perspective, *Nemawashi* or the theory of 'root binding' by a process of consensus and conflict, is a Weberian construct describing the Japanese bureaucracy as an efficient policy-making machine, maximizing its information-gathering proficiency through a vast organizational network. *Nemawashi* also has a cybernetic feature since the same information lines that are vital in accumulating data serve as two-way communication channels to control public opinion and forge consensual behaviour among important policy sectors. The third perspective, *Nawabari Arasoi* or the bureaucratic politics approach, is a description of bureaucratic activity being directed and regulated by incentives for budget increases, expansion of departmental programmes and control over manpower resources.

These viewpoints are not seen as mutually exclusive alternatives, but as contemporary theories describing different social aspects of policy formation in Japan. What one view can contribute to illuminating the complexities of the actual policy-making reality may be considerable, but is limited. The *Nebukai* argument about the close working relations between different sectors, for example, fails to explain strained interactions often seen in *Nawabari Arasoi* between private enterprises and government, or between ministries fighting for control over specific policy areas in biotechnology.

For this reason, an adequate understanding of Japanese decision-making behaviour in biotechnology requires a critical synthesis of these three perspectives into a unitary *Nebukai–Nemawashi–Nawabari Arasoi* synthesis, what is referred to throughout as the Triple N Synthesis. Only from this standpoint can there be a multilateral analysis of all the empirical evidence. In the process of reaching this final picture, the basic structural elements underlining the policy formation procedure and the fundamental issues confronting the policy-makers are assessed, and factored into the overall conclusion.

What has been difficult to account for in the past is the dynamism and flexibility often found in the Japanese economy. The flexibility in the fact that many presently structurally depressed industries were two decades ago 'targeted' areas of economic growth. The attempts of the state-led capitalist school to offer explanations for this dynamism have begun with assumptions of the bureaucracy harbouring a unique internal mechanism. This bureaucratic structure was thought peculiarly Japanese since in most countries state-led economies are expected to be slow and non-adaptive like the lacklustre bureaucracies that oversee them. Bureaucracies as organizations are notorious for gradual adjustment to changing environments, and only exogenous factors are supposed to explain their transformations. Michael Crozier suggests, 'A bureaucratic system will resist change as long as it can; it will move only when serious dysfunctions develop and no other alternatives remain.'[1] Consequently, change in a bureaucracy is difficult to enforce requiring necessary long delays, much determination in overcoming a strong resistance, and large efforts to revamp the entire ministerial operation. From this viewpoint, the Japanese example was difficult to comprehend.

The '*laissez-faire* economist' school attributed Japanese dynamism to the spirit of competition which stimulates Japanese companies to promote technological innovation and economic prosperity.[2] Of the 148 members listed in 1986 by a MITI-affiliated think tank as participating in biotechnological processes, for example, over fifty Japanese firms were competing to produce a limited variety of monoclonal antibodies for medical diagnostic kits and as laboratory tools.[3]

In proposing the Triple N Synthesis of Japanese political behaviour in biotechnology policy development, we are attempting to elucidate the central point of ambiguity in the overlap of the 'state-led capitalism' and the *'laissez-faire* market forces' theories; namely, that the Japanese postwar industrial policy success is based on a constantly evolving interaction between the public and private sectors. Any attempt to reveal the nature of the dynamism and flexibility in the Japanese economy, therefore, can only partially be successful by examining standard structural components in either specific bureaucratic practices or in the nature of a firm. Alternative methods of financing by MITI, or political structural elements such as the insulation of the Japanese bureaucracy from powerful interest groups, do undeniably give flexibility. The same could be argued for the competitive nature of the Japanese firm, or for its particular practice of company-based labour unions. But a full understanding of the biotechnology case can only be achieved by considering the structural features of a firm or bureaucracy as well as an analysis of the dynamism and flexibility in their complex public–private sector interactions. In fact, all three components of the Triple N Synthesis share the capacity to explain, albeit from different perspectives, the active relationship of the bureaucracy with other actors or with itself in shaping biotechnology policy.

If this dynamic interaction of policy actors is so important, then the Japanese biotechnology policy environment must not only demonstrate constructive exchanges between policy participants, but also should be replete with examples of intermediary structures both formal and informal that facilitate these exchanges. Does the Japanese biotechnology case give clear evidence of the nature and function of these bodies? If we assume that dynamic interplay has been a cornerstone for postwar Japanese industrial policy, then we must also ask how relevant past industrial policies are to that of Japanese biotechnology. Are the same mechanisms of previous successes, or perhaps cosmetic reconsiderations, still dramatically influential? Will Japan be as economically successful with biotechnology as it was with shipbuilding, cars, consumer electronics, and so forth? This kind of questioning leads directly to further inquiries about the effectiveness of biotechnology policy or at least, its estimated effectiveness for the future of the Japanese economy; a theme implied throughout this book.

4

Chapter one

Understanding Japanese policy-making with the 'Triple N Synthesis'

Over the past few decades, an attentive public around the world has developed, in a very broad sense, a convergence of views about Japan. Today, unlike thirty years ago, this small Pacific archipelago of 120 million people is seen as an industrial Midas sweeping world markets in almost every sector it 'touches'. In fact, like the gift of the mythical king himself, the success Japan has earned selling cars, video recorders, televisions, cameras, and the like, has in recent years begun to haunt it with threats of trade wars, rising foreign protectionism and a new skepticism for 'things Japanese'.

Much of both the criticism and admiration for the Japanese by Western observers has focused on either the role of the government in managing the economy, or the competitive nature of the Japanese private firm. Earlier accusations of Japanese protectionism against huge multinationals dominating her small yet bustling domestic private sector were epitomized by 'Japan, Inc'; a conspiratorial plot of government and big business towards non-Japanese enterprises.[1] Today, the US government, in particular, often insists that the Japanese have substituted outright protectionist measures with equally insurmountable non-tariff trade barriers to insulate domestic businesses from the harsh realities of foreign competition at home. On the other hand, however, other foreign governments view the Japanese as masters of reallocating scarce resources to new business areas that have been 'targeted' by the central bureaucracy.

Targeting in Japan is described by the US International Trade Commission (USITC) in a 1983 report as, 'Coordinated government actions that direct productive resources to give domestic producers in selected industries a competitive advantage.'[2] Daly adds that for biotechnology policy, 'Such actions include the development of infrastructure, the adjustment of tax, patent, export and regulatory laws, direct investment in industry, provision

5

of venture capital, and the encouragement of industry consortia to maximize economies of scale.'[3] In fact, Japanese government officials themselves have adopted the English, and often refer to '*tarugetto*' of an industrial sector. Allen argued that 'Much of Japan's success so far has been associated with her choice of industries for development in the early period of her postwar career'.[4]

The state's 'active role' in the development of national priority sectors in Japan has been a topical issue amoung Western 'Japanologists'. Exaggerated foreign media coverage and casual observers have often reduced the role of government intervention in the Japanese economy to the virtues of one bureaucratic organization in particular, the Ministry of International Trade and Industry (MITI). It is no surprise, therefore, that when MITI announced a commitment to electronics, biotechnology and new materials in its April 1980 *Vision for the 1980s,* and subsequently supplemented it in the following year with a national project for 'targeting' these three areas, there was an international over-reaction. Traditionally in Japanese postwar politics, these long-term plans or visions were meant to provoke domestic firms into redirecting investments – what Stockwin describes as 'an announce-ment effect.[5] But this 'announcement effect' gained a broader regional significance, sending shock waves to all corners of the world economy. Articles about the coming Japanese onslaught in biotechnology-related products were abundant, featuring prom-inently in popular magazines, local presses, and even academic journals.

The most authoritative statement to date, and perhaps the one that precipitated the most international feedback, was the report on biotechnology in January 1984 by the Office of Technology Assessment (OTA) of the US Congress. This widely read document concluded that although the US is pre-eminent in biotechnology, 'Japan will be the most serious competitor of the United States in the commercialization of biotechnology.'[6] The OTA, which often purports to be Congress's private think tank, suggested that the Japanese strength stemmed from their long involvement with 'bioprocess engineering' in their traditional fermentation industry, and the tremendous potential of their domestic pharmaceutical market, one of the first industries thought to be important in the future application of biotech-nology. The report recognized and stressed the significance of a general consistency in Japanese industrial policy which seems to operate whenever that nation becomes involved in a new in-dustry. This, they argued, is due to myriad companies launching

6

desperate struggles to develop the same products under the co-ordinating power (*chōseiryoku*) of MITI. The report under-lined MITI's talent in harnessing some of this competitive energy in large co-operative national programmes.

Despite some inaccuracies, such as the lauding of a moribund project (The Next Generation Basic Technology Project), the OTA report justifiably created a social awareness in the world of the importance of biotechnology to future international productivity. It also managed, as a corollary, to provide an understanding of Japan's lack of hesitation and long-term com-mitment in exploiting the rich potentials of this new field.

The immediate effect of the OTA assessment was a backlash reaction criticizing the lack of 'targeting' in developmental strategies for biotechnology in the West. With persistent references to the Japanese, American and European scientists and business leaders, in particular, called for bigger commitments to biotechnology from their governments. George Ruthmann's comments epito-mize the sentiment, 'Although I do not think there is any way to quantify it, it would appear that the Japanese commitment far exceeds anything going on in the United States.'[7]

The cries for more government participation were less pleas to adopt a Japanese policy decision-making style by means of increased centralization or co-ordination, as they were desperate calls for financial help shouted out of an anxiety that the Japanese were better prepared for the future. The feeling was that the West somehow lacked an essential element in the absence of a long-term government commitment. Even US firms, usually content with deregulation and decentralization of policy decision-making, seemed to suggest that a nationally defined target for promoting excellence in biotechnology would serve as a 'government guarantee' of future commitment. This reassurance from government was needed, it was argued, to isolate American business against the imminent competition of Japanese bio-technology companies.[8] This kind of appraisal of Japanese strategies in implementing policy has led to an exaggeration of the efficacy of the Japanese bureaucracy's policy instruments in promoting specific industrial sectors. Ronald Cape, the president of Cetus corporation – a small biotechnology firm that has grown considerably on venture capital – suggested Japanese style 'biotechnology institutes' to 'bring together academic and industry investigators for cross-disciplinary research activities'.[9] If these 'biotechnology institutes' refer to Japanese research consortia (the only government-sponsored 'institutions' with academic and industrial researchers), others argue that the policy suggested for

7

the US have serious anti-competitive effects, and would strengthen the oligopolistic nature of an industry effectively excluding small firms, like Cetus corporation, from large profits.[10] Nevertheless, the concept has gained some favour, as seen in the UK, where the recent establishment of enterprise 'clubs' for joint company research hints of some Japanese formative influence.

Despite enthusiasm abroad, in Japan most official ministerial reports about the status of biotechnology convey a sense that Japan is, in general, trailing Western nations. MITI's Bioindustry Advisory Council (BAC) recognized Japan's traditional strengths in fermentation in their influential 1984 report, but emphasized her weaknesses in the new biotechnology; 'Japan, having nurtured traditional fermentation and distilling industries, has the world's best technology when it comes to the industrial application of microorganisms ... However, Japan is weak in basic and applied research and in new fields, like rDNA technology, she is generally behind (the West)'.[11]

Like most Japanese government reports, the MITI document invariably judges weakness by the development of a particular area relative to its degree of sophistication in Western nations. In discussions about information resources in biotechnology, for example, the MITI-sponsored committee warned that, 'At present, Japan's system is several years behind that of the US and Europe, making it necessary to plan the rapid construction of a system (in Japan)'.[12]

In sharp contrast to this pessimism of government, the Japanese private sector has shown increasing signs of optimism *vis-à-vis* its standing with the West. Dore has indicated a striking change in the self-confidence of many Japanese firms instigated by the successful growth of the electronics industry in the 1970s.[13] New liberalization measures that private enterprises initially feared would expose them to the harsh realities of international competition, instead have helped reinforce their realization of the quality of Japanese products, their substantial distribution network into foreign markets, and their considerable experience in monitoring the importance of sophisticated technology abroad.[14] According to a survey of the research and development (R&D) activities of private companies conducted by the Science and Technology Agency (STA) in April 1984, over three-quarters of the companies questioned thought that they had attained aggregate, technological levels either as internationally competitive or superior to comparable firms in the US and Europe.[15]

There is, therefore, a mixed bag of perceptions, both

8

internationally and in Japan itself, about the potential of biotechnology-based industry. First, in the industrialized nations of the West, the Japanese centralized, goal-oriented strategy of policy initiatives towards biotechnology has incited fear in both the public and private sectors. This fear articulates the notion that it is only a matter of time before Japan reaps the fruits of a world pre-eminence in biotechnology. Second, the Japanese private sector, although a great deal more cautious in its predictions than outside observers, has agreed with this assessment. It is striving to achieve a critical level of biotechnological capability enabling large increases in industrial growth and productivity. Third, in direct contrast, the Japanese government does not share this ease of drawing conclusions from general propositions, and cites Japan's late start, its lack of basic researchers, and an underdeveloped supporting technological infrastructure, such as scarce information data bank networks, as critically disadvantaging Japan in biotechnology.

Given these contrasting opinions about the efficacy and nature of Japanese efforts in biotechnology, how is it possible to assess Japan's capability in this field, and in particular, the essence of the role played by the Japanese government? This book offers the Triple N Synthesis as an approach to understanding the complex processes in Japanese policy-making in biotechnology and the functions of the important policy components involved. This combined approach in part asks, 'Can Japan become a world leader in biotechnology-based industry?' 'Are there adequate similarities between the way the nation is pursuing policy now compared to the 1950s and 1960s, and if not, what are the differences and how effective on biotechnology have they been?' The significance of the Triple N Synthesis – a full exposition of which is left to the final chapter – lies in its analytical power as a synthetic theory composed of three structural components that interact with each other.

The Triple N Synthesis

Already there are many models and approaches that analyse policy-making in Japan. Nevertheless, conceptual guidelines are useful in defining questions, facilitating discussion and offering plausible explanations for policy outcomes. The Triple N Synthesis approach is offered to search for examples of structural dynamism in Japanese policy decision-making and implementation practices in biotechnology and to guide us through the pitfalls of single-factor explanations.

9

In order to make a more fruitful discussion of the composite picture it paints, a distinction must first be made between 'policy-making' and 'decision-making'. The former, as used in the title to this chapter, is the overall process of policy action involving decision-making and decision implementation. Decision-making, on the other hand, implies a single, critical choice from a set of preferred alternatives. Rarely does policy-making hinge on one decision, but is rather a series of decisions in a well-defined process.[16]

The three conceptual frameworks comprising the Triple N Synthesis are:

1. *Nebukai* – viewing policy as a result of a cohesive, conservative coalition
2. *Nemawashi* – as an outcome of concensus and conflict
3. *Nawabari Arasoi* – as the final achievement in the game of bureaucratic politics

The Triple N Synthesis describing Japanese policy-making in biotechnology is an elitist view proposing that biotechnology policy is a product of a special few formulated according to their needs and values. The working assumption is that the values of Japan's elite – academic researchers, business leaders, civil servants, and LDP politicians – soon become the values of society at large through an effective mass media and communication network. The extent of the public receptivity of the government's policies is demonstrated by public acceptance of biotechnology products as self-evidently safe, and the lack of national dissension as epitomized in the US by the environmentalist, Jeremy Rifkin.

There is no 'best' perspective of the three views in the Triple N Synthesis that adequately explains Japan's biotechnology process all at once, since there is no unicausal explanation for a country's political phenomena. Each perspective instead focuses on different aspects of policy-making given the same empirical data, and in certain cases, one proves more suitable than the others at explaining specific policy outcomes.

It is useful to remember that the 'real world' scenario is far more complex, and that the three separate perspectives are only different 'windows' (*madoguchi*) or 'lenses' from which the agglutinated reality can be partly deciphered.

Nebukai – cohesive conservative coalition approach

The *Nebukai* perspective of the Triple N Synthesis simply hypothesizes that in the policy formation and implementation of biotechnology, the ethos of the Japanese government (the state and the ruling party) is very similar to that of the private sector. *Nebukai* reduces all its reasons for close, structural cohesion between the ruling party, the bureaucracy, and big business to a theory of historical evolution. The tradition of experience in postwar industrial development in Japan, it is argued, has created a normative, regular set of relationships between these three sectors. The word, *Nebukai* in Japanese literally means 'deeply rooted', and refers in this case to the long-standing and deeply consolidated working relationships that have evolved between the three sectors. It is a perspective that holds the development of biotechnology policy-making as a larger consequence of a postwar policy-making pattern that has changed little over the years.

In analysing the interpenetration between the actors of the *Nebukai* view, each of the three relationships is taken in turn and examined below.

Political–bureaucratic relations

The primary reason offered for this close bond is the stability created by the long-standing, unbroken rule of the Liberal Democratic Party (LDP). In fact, the domination of the LDP has created a *de facto* political partnership between a conservative party and an elite bureaucracy. Since the November 1955 formation of the LDP, this partnership has developed, over time, a certain permanence, and has been moulded by successive policy issues and societal changes into a fluent, flexible coalition.

Several reasons exist for the success of this synergistic partnership. First, the absence of any deep ideological cleavages between the two political institutions. Second, the long-tenure of office by the LDP has produced a clear distinction of roles between the two sectors: such as the prevalence of the ministries in economic policy and the influence of the politicians in education. Third, the bureaucracy has been fairly successful in stimulating healthy postwar economic growth rates which have raised standards of living for most parts of Japan, and contributed to the consolidation of the LDP as the dominant political party. Fourth, a weak and fragmented opposition has left the bureaucracy with sufficient autonomy to develop policies in a relatively sheltered, unchallenged environment. Fifth, the mutual interpenetration of the two sectors is marked by both formal and informal ties, such as ex-bureaucrats joining the LDP, advisory

11

Biotechnology in Japan

counselling of politicians by bureaucratic leaders, joint advisory councils, old school ties, and so on.

It is to be demonstrated that the development and implementation of biotechnology policy depends heavily on the strength of this bond.

Bureaucracy–private sector relations

The *Nebukai* view asserts that state-private sector co-operation is also a result of a long historical process. Less-developed capital markets, more susceptibility to state-sanctioned policy measures, and large-scale borrowing from state incorporated banks have been integral to the dependence of large firms on state intervention. The longer these two sectors co-operate effectively on political and economic issues, participate in joint R&D financing, and cultivate information exchange, the more deeply fixed is their mutual trust and interdependence. In the resolution of the environmental pollution disasters of the 1960s and early 1970s, for example, even though the courts held the private sector ultimately responsible, the civil service also accepted blame for the careless scourges of companies such as Chisso and Showa Denko.[17] For the Japanese public, the cohesiveness of the *Nebukai* coalition already meant identifying private sector polluting with bureaucratic and LDP foot-dragging in the latter's reluctance to address the former's behaviour.

The *Nebukai* view hypothesizes that throughout the postwar period, the roles of definite organizational entities such as public corporations, special non-profit institutions, information gathering think tanks, big business and industrial federations, and the like, have been very important in this close interaction between the bureaucracy and large industrial firms. It can scarcely be overstated how other factors such as informal and formal networks (*jinmyaku*), *Amakudari* (ex-bureaucrats landing top administration jobs in the private sector), and old boy ties have also contributed to the shared proximity between the two sectors.

With the absence of the military, the *Nebukai* coalition has ensured that the state is able to devote most of its economic strength to the private sector. As Okimoto asserts, the efficiency of Japanese industrial policy is perhaps only in its cost-effectiveness in ensuring optimal use of precious national R&D resources.[18] *Nebukai* predicts similar patterns of close private–public sector relations with biotechnology.

12

Political–private sector relations

As with political–bureaucratic relations, the long rule of the LDP is a large contributing factor to the steadfastness of this bond. Particularly, it is responsible for the favourable and prosperous environment enjoyed by most in the private sector. The successful relations between the ruling party and industrial giants are marked by the LDP's pro-business orientation and the broad base of support for the ruling party nurtured by big and small businesses alike. In part, the cohesiveness of their relationship is at the expense of organized labour. It is no secret that both the LDP and private companies in Japan mutually benefit from the weakness of nationally organized labour.

The united coalition

The resultant tripartite *Nebukai* coalition formed between the LDP, the bureaucracy, and the private sector is necessarily conservative and powerful. The *Nebukai* view portrays the three sectors functioning as a unitary decision-maker in a well-co-ordinated team. This implies a team leader to ensure single-mindedness of effort and unity in timing of members' actions. The state satisfies these criteria, and as the dominant coalition member, is responsible for central co-ordination and control of the other two members.

As a result of the evolution of these close relationships and the co-ordination in timing in implementing policy strategies, the *Nebukai* coalition predicts a 'follow-the-leader' behavioural pattern among big businesses with regard to the state. At the same time, the bureaucracy would be expected to lend a quick ear to the recommendations, advice, and suggestions of either big business or politicians.

The word 'coalition' also suggests balance and stability in the policy-making machinery; a functional triumvirate matured through 30 years of policy experience. The undesirable by-product of this cohesion between members is that the coalition's actions are in danger of being interpreted as too nationalistic: deliberate moves by Japan towards one strategic objective – economic advantage over the rest of the world.

The dominant coalition perspective is the most vulnerable of the three components of the Triple N Synthesis to be exaggerated and misinterpreted by casual outside observers. It has several important distinctions from the infamous 'Japan, Inc.' metaphor, which describes state-industrial relations as the familiar, almost familial sociopolitical environment of a huge industrial firm. First, the term 'cohesive, conservative coalition' is not pejorative like 'Japan, Inc.' in the sense of connoting a premeditated plot to insulate all Japanese industry from foreign competition, no matter the cost. Instead, it points to the merits of co-operation and effective co-ordination in discussing policy matters within a country. Second, *Nebukai*, although accentuating co-operation, implies more of a heated bargaining relation among coalition members than the outright exclusion of conflict implicit in 'Japan, Inc.'. (To the extent to which this first view, *Nebukai* overstates 'cohesion', the third approach *Nawabari Arasoi* exaggerates 'conflict'.)

The national interest sentiment of *Nebukai*, however, is conversely one of its main strengths. Three coalition members sharing common goals of working for the collective good creates a domestic policy environment that is generally conducive to consensual decisions, and an overall support of the final outcome. The nation, in this sense, is the 'ultimate collectivity'.[19] During the era of high economic growth, the collective interest consistently led to a consensus on the importance of yielding large economic dividends for the private sector. Later in the 1970s, this value system changed as environmental pollution, a by-product of the formal 'economic-growth-for-economic-growth's-sake' policy, became a critical issue. In the 1980s, no longer is optimal economic activity at all costs the foremost national policy objective. MITI's *Vision of the 1980s* was adamant in its support of enhancing the quality of Japanese daily life (see Chapter Three).

In summary, *Nebukai* describes Japan's policy decision-making and implementation machinery as a coalition of three distinct members – the bureaucracy, big business, and the LDP. Decisions are centrally controlled, value-maximizing, and based on the best information possible. The process assumes clear objectives of national goals, systematic ranking of alternatives, and the three-member coalition selecting options as a unitary, rational decision-maker.

Nemawashi – consensus and conflict in policy-making

Nemawashi unveils the magical black box of the cohesive coalition, and further reduces the broadly defined groups of the coalition – bureaucracy, big business, and the private sector – into smaller, more analytically convenient units. The *Nemawashi* perspective critically reveals the many participants outside the triumvirate who have influenced biotechnology policy.

Nemawashi is literally 'root binding' in Japanese, and refers to laying the groundwork for obtaining a broad agreement about policy objectives among a wide consortium of players. Laying groundwork usually involves lengthy consultations with many concerned actors; negotiations in which conflict, disagreement and dissension are a natural, perhaps necessary, part of achieving a final broad concensus.[20] In applying *Nemawashi* to the Japanese bureaucracy, a small, yet important distinction must be made between 'root binding' among the umbrella organizations of a single ministry, such as its advisory bodies (ABs), or related auxiliary groups (*gaikaku dantai*), and 'root binding' among a collection of ministries each with the same policy objective. The former is discussed here, whereas the latter is the basis for *Nawabari Arasoi* described later.

Nemawashi, unlike the *Nebukai* coalition perspective, identifies the multiplicity of persons, groups, and organizations who participate in the daily policy decision-making and implementation processes. Applied to the biotechnology policy case, *Nemawashi* predicts that academic researchers, both those connected with the major public universities as well as those with the national research institutes, are important policy participants overlooked by the *Nebukai* coalition analysis. A second prediction is that the amorphous collective of the 'bureaucracy', as defined in *Nebukai*, refers to the activities of ministerial units such as MITI, the STA, the Ministry of Health and Welfare (MHW) and the Ministry of Agriculture, Forestries and Fisheries (MAFF), as well as to 'think tank' organizations loosely affiliated with these ministries. Again, in the political sphere, this analytical approach gives an added sense of pluralism by highlighting not only the LDP, but opposition parties involved. In biotechnology policy this would mean the identification of the *Kōmeitō* (the Clean Government Party), and its advisory body on biotechnology. Furthermore, *Nemawashi* reduces the all encompassing term, 'big business', as used in *Nebukai* to its analytical units. This means that the *Nemawashi* perspective is perhaps useful in pinpointing policy inputs by *Keidanren*, known in English as the Japanese Federation of Economic Organizations, as well as from individual large firms. In addition, most importantly this analytical view of Japanese policy decision-making considers the influence of those players most displeased with promoting biotechnology. These mainly include local opposition groups who protest primarily on ideological grounds. Finally, *Nemawashi* postulates that in bio-technology policy, the same information and peripheral contacts vital to the operation of any organization, for example MITI, can

15

be used simultaneously to disseminate information in a cybernetic manner. In practical terms, this means the creation of a consensus-building network for each ministry to affect those under its umbrella.

The *Nemawashi* approach is a variant of a cybernetic-organizational process model, since it assumes that the analytic unit for discussion is the output of organizations, that policy decisions involve interactions between 'a conglomerate of semi-feudal, loosely allied organizations, each with a substantial life of its own', and that policy implementation requires mass consensus building through the communication channels between organizations.[21] It does not refer to a monolithic coalition functioning to promote the attainment of a collective goal, but denotes instead a set of institutions administered by human beings following certain priorities and standard operating procedures (SOPs). Under this rubric, even the opposition groups protesting against biotechnology can be described as being united in an organizational structure (see Chapter five). Structural constraints aside, policy strategies formulated depend on the interaction of people or at least small groups within the organization. Expectations of policy are based on available information, and choices of specific action on the recognition of the first available alternative thought acceptable for the goals pursued.[22]

SOPs are fixed rules established in an existing organization prior to a particular instance of policy action.[23] The constraints imposed by conventional organizational practices ensure a consistency of behaviour both for the policy outcomes of the organization as a whole, and for the people interacting inside it. It makes little difference how the SOPs came to be; the possibility, for example, that they are evolutionary products of the cohesive coalition construct is interesting, but of little concern. What matters is that these SOPs exist as constraints on policy outcomes, and if the large numbers of people working in an organization are going to be effectively co-ordinated, the SOPs need to be followed.

One of the classic works for reference in analysing outputs produced by organizations is March and Simon's *Organisations*.[24] March and Simon maintain that human beings as problem solvers in large organizations are constrained by certain limitations, namely a finite physical and mental capacity to generate novel strategies, process fresh information, and solve new problems. Focusing on the organizational outcomes produced by the imperfect human being, they propose several characteristics of organizational action.[25]

16

1. *Factored problems* – Some problems are so complex and difficult to resolve that they are best 'factored' out into smaller, more manageable parts and given to specialized subunits. In biotechnology, this would mean that a ministry, such as MITI, handles most of its difficult problems by assigning small tasks with subgoals suitable for an organizational subunit – a kind of administrative division of labour. The abundance of advisory bodies and affiliated organizations (*gaikaku dantai*) of many bureaucracies are *de facto* subunits for sharing the burden of a policy problem.

2. *Satisficing* – Humans in organizations do not work by constantly searching for the most desirable alternative to a problem, but simply cease looking if a satisfactory option has been found. For an organization, finding a needle in a haystack is better than searching for the sharpest needle in a haystack.[26] Related to satisficing is the idea that the order of alternatives is crucial since in a search for the most suitable alternative, the first one found is the option adopted.

3. *Risk avoidance* – People in organizations are reluctant to take gambles in policy decisions based on uncertain estimates of the future. A reliance thus develops on methods giving short-term feedback of policy and fresh information to correct deviations from desired outcomes. If this is so, then in biotechnology we would expect advisory bodies and information gathering subunits to be associated not only with the ministries, but also with the national research institutes, *Keidanren* as well as the LDP.

4. *Repertoires* – In recurring policy situations, organizations tend to make the same efficient choices as they have made previously. In Japanese biotechnology policy, this would mean ministries are very quick to copy existing action programmes and national projects adopted elsewhere. Similarly, ministries, especially MITI, probably employ those same policy tools in biotechnology that were used in consumer electronics, cars, and shipbuilding. We would also expect to find Japanese biotechnology policy mirroring that of America in many areas. The prevailing attitude is expected to be that if the policy has proved successful in the US, why not employ the same practice to similar problems in Japan.

Bureaucracy in the Nemawashi *perspective*

Since *Nemawashi* hypothesizes that government ministries and agencies are primarily the locales where the final say rests in both Japanese biotechnology policy decision-making and implementation, it is important to consider the essence of the bureaucracy as an organization. In the literature about organizational theories of bureaucracies, there are conflicting ideas in discussions of how a bureaucracy works. On the one hand, starting with Max Weber's work on the 'ideal type' of bureaucratic organization, a picture has been sketched in the last fifty years of a rational, efficient, highly specialized group whose actions have far-reaching implications for modern society.[27] On the other hand, a contrasting analysis is suggested, sometimes by the same authors, of a bureaucracy as a giant leviathan constrained by too much centralization, inefficiency, structural rigidity, and the institution within of an 'iron law'.[28] Although these views straddle the realities of the Japanese bureaucracy, they are useful as rough parameters to develop a basis for understanding the structural constraints involved.

In Japan, individual ministries and agencies as organizations have unique traits *vis-à-vis* biotechnology policy that distinguish them from non-central government organizations. First, the evolved role of central government organizations carries an obligation of public accountability. This feeling of serving the public is usually expressed in each ministry by a strong organizational identity in addition to a sense of intense nationalism. Second, Japanese bureaucratic organizations are highly specialized; manned by an educated elite who often share the same university degrees and similar scientific skills as those for whom the policy is being made. Third, superior human resources in the Japanese bureaucracy are paralleled by excellent material resources, such as a constant feedback of information from a wide network of reliable sources and lucrative sources of funding. These differences in organizational make-up between non-government organizations and bureaucracies, it is argued, have resulted in Japanese ministries and agencies being allocated more power for general policy decision-making and implementation than non-government institutions.

For these reasons, *Nemawashi* in Japan must inevitably involve one or more ministerial organization in the bargaining games between the various groups. It is a testimony to the Japanese political structure that the power in the bureaucracy does not become self-absorptive and oppressive, as in some nations with a strong state. Instead power is shared in part through an organizational system of bargaining with a multiplicity of actors. It is ironic that the nature of the *Nemawashi* perspective demands that the very networks through which information is derived must be used to promote policy implementation and consensus building. This means that information networks are in essence two-way communication channels for retrieving and disseminating data. Primary among the channels by which information is gathered, especially in the Japanese bureaucracy, is the advisory bodies (ABs). It is one of the hypotheses of the *Nemawashi* analysis that ABs used for consultation and advice on policy issues are also keys to effective consensus building and the coordinating power of the Japanese bureaucracy.

Nemawashi advisory bodies

The system of employing advisory bodies (ABs), especially in the Japanese bureaucracy, 'to listen to the opinions of those concerned administrators in the business and academic world, debate the issues, consider policy direction, demands and propositions and formulate a national consensus,'[29] has become institutionalized in recent years. The rise in popularity of ABs is often attributed to their frequent use by Premiers Ohira and Nakasone.[30] Nakasone, especially, established *Shingikai* to examine a variety of problems. These included a private *Kondankai* (Committee) to the Chief Cabinet Secretary to investigate the politial uproar over the Cabinet's visit to the Yasukuni shrine, a private *kenkyūkai* (Committee) to the Prime Minister to report on raising the one per cent ceiling in defence spending and the Temporary Investigation Council on Education (*Kyōiku Rinchō*) to study the revamping of the educational system. The use of ABs in the political and civil service sectors has encouraged the private sector, national research laboratories as well as ministerial auxiliary organizations to adopt ABs to solve pressing policy problems.

It is difficult to discuss in absolute terms the precise codification of ABs since they are so numerous, and the territory they cover broad and relatively uncharted. As a general rule, however, ABs in Japan can be divided into either a statutory or non-statutory category. Statutory ABs are widely known as *Shingikai* (often translated 'Council' in English), but also refer with some exceptions to *chōsakai, shinsakai, bukai, iinkai, kyōgikai,* and *kaigi* (usually all translated 'Committee'). They number approximately two hundred in the Japanese civil service.[31] Many of these statutory ABs are called 'public advisory bodies' *(kōmonteki shimon kikan)* alluding vaguely to their organizational goals of serving the public good and opening the results of their reports to the general populace. Although non-statutory ABs have no generic counterpart of *Shingikai*, the word *Kondankai* is closest as a functional equivalent. In this book, *Kondankai* refers to *ad hoc* organizations known as *kondankai, kenkyūkai, warukingu guruupu, bunkakai, konwakai* and *benkyōkai* (translated both as 'Committee' or 'Subcommittee'). Since these are informally established ABs, no one is certain of the exact number, although a survey of the daily papers from January 1984 to September 1985 yielded nearly three hundred.[32] A large proportion of these non-statutory ABs are called 'private advisory bodies' *(shiteki shimon kikan)*, offering specific advice usually to a bureau chief or even to a minister, and are specifically for use, at least initially, within a particular bureaucracy.

Although both statutory and non-statutory ABs are crisscrossed by 'permanent' and *'ad hoc'* committee groupings, as a general rule, statutory ABs can be though of as 'permanent' and non-statutory as *'ad hoc'*. This distinction is not absolute throughout the entire Japanese polity, and is offered here only as a rough conceptual outline to understand the role of ABs in Japan. It is the classification of an AB as either permanent or *ad hoc* rather than its legal status that determines the importance of its function.

Permanent ABs are often influential organizations that have become intitutionalized in their own right as an important part of the policy decision-making structure of a bureaucracy. They are usually responsible for overall strategies or are referred by officials in the ministry to address large policy problems. With regards to science policy, for example, each ministry has a science *Shingikai* – with its own SOPs, regularized channels for reaching decisions and formal information networks – that is very influential in the planning and implementation of strategy in science. Two of these *Shingikai* in particular, the Council of Science and Technology (CST) of the Prime Minister's Office – also affiliated with the STA – and MITI's Industrial Structure Council, have considerable influence. The *Nemawashi* perspective hypothesizes that these science councils play crucial roles in collecting information for the bureaucracy in the biotechnology policy decision-making process.

Ad hoc ABs are temporary committees established to handle a specific policy problem and are quickly disbanded once their task is completed – usually after the submission of a report. *Kondankai* can be divided functionally into two groups. The first group has a role similar to that of a discussion roundtable. A *Kondankai* organized to discuss the viability of a research project, for example, would have tasks such as establishing a research theme, forming a suitable budget, and publishing a report of collected opinions about the proposed project. In this capacity, *ad hoc Kondankai* are provocative fora for the critical debate of hotly contended issues. After opinions are aired and a working agreement on prior differences printed in a report, the temporary *Kondankai* are usually disbanded.

If, however, the issues debated grow progressively more important over time, the *Kondankai* shed their temporary status, and undergo 'developmental restructuring' (*hattenteki ni kaiso*) or 'progressive dissolution' (*hattenteki ni kaishō*) to form permanent AB committees, known as an *Iinkai* (Committees). In this second case, the original *Kondankai* are referred to as *botai* (literally, 'mother organizations') – foundations from which the more prestigious, permanent *Iinkai* (Committees) are created.

The classification of *Shingikai* and *Kondankai* as permanent and non-permanent ABs respectively often confuses Western observers, especially when they are both referred to in official English language government translations as, 'Committees'. Typically, a *Kondankai*, such as *Keidanren*'s Life Sciences Committee, remains a *Kondankai* for several years and, for all intents and purposes, seems to be a permanent AB. Then suddenly, a new-found importance in the life sciences forces a promotion in the committee's status, and hence a change in name to an *Iinkai* Committee. In order to understand these subtle differences in a 'committee's' role, both *Shingikai* and *Kondankai* ABs are best referred to by both their Japanese and English names; for example the 'Life Sciences *Kondankai* Committee', the 'Bioindustry Advisory *Iinkai* Committee', or the 1980 MITI Vision *Bunkakai* Subcommittees.

Most ABs demonstrate diversity among committee members; a pluralism of policy inputs with the ultimate objective being the consolidation of a consensus among a large group of players. It is popular among Japanese intellectuals to criticize central government ABs as mere tools of credibility for ministries. ABs, they argue are only 'fronts' (*tatemae*) for advising on policy, and in fact, have no real significance except to endorse proposals created by ministerial officers. They maintain that despite measures during the Occupation to maintain the independence of ABs, ministries have used them as instruments of power or disseminators of information in order to control policy direction and to expand available resources.[33] 'Properly' functioning ABs, it is argued, should on occasion be able to forsake policies advocated by particular ministries and not merely rubber stamp already planned and sealed policy proposals from the government.[34] Criticisms are aired of individual ministries handpicking ideologically 'safe' committee members – usually academics – and many refer to lessons learned in science policy from the marginalized Japan Science Council (JSC) (*Kagaku Gijutsu Kaigi*), a permanent AB of the Ministry of Education (MOE) that fell into disfavour because of its advocacy of 'an independent political line'.[35]

MITI is also thought to have learned from repeated occurrences of allowing *Shingikai* to behave with too much autonomy and independence immediately after the Second World War. The situation came to a head with the formation of a private AB to the Minister of MITI, the '*Shingikai* to Reorganize the Electricity Utilities'. This *Shingikai* was established in January 1949 to discuss the break up of the Japan Electric Power Generating Co., Ltd and the future of nine large regional electricity distributing companies run jointly by government and the private sector. The Japan Electricity Power Generating Company had held a monopoly on the production of electricity since its formation in 1939.[36] Its dissolution was imperative because of the Occupation's 'Law for the Elimination of the Concentration of Excessive Economic Power'. The four-member *Shingikai* was formed after a previous 19-member committee of businessmen, the Committee for the Democratization of the Electricity Utilities, failed to produce a proposal for dissolution suitable to General Headquarters (GHQ).

The controversy developed when MITI, pushed by the Americans, suggested to the *Shingikai* in May 1949 that seven smaller companies be formed from the monopoly. The appointed Chairman, Matsunaga Yasuzaemon, nicknamed the 'Electricity Devil' (which gives the connotation in Japanese – *Denryoku no Oni* – of one who is tough and extremely competent), believed in the freedom and autonomy of both the *Shingikai* and its members in debating and exchanging ideas. In January 1950, the Matsunaga Plan was presented for the dissolution of the large Japan Electricity Power Generating Co., into nine regional companies capable of both generating and distributing electric power.[37] The proposed plan was contrary to MITI's wishes and instead supported those of the electricity companies. MITI was furious and in April 1950 the plan was rejected by the Diet. The GHQ intervened in July by announcing the discontinuation of funding to the electricity companies as they stood until a Reorganization Law was passed. MITI had no choice but to relent, under time pressure from GHQ, to Matsunaga's wishes and, after much bitter debate, the original plan was announced by MacArthur as law on 24 November 1950.[38]

Since the ministerial officials did not even attend the *Shingikai* meetings, they had no bases for revising the final recommendation of the committee which gave rise to the nine existing companies in operation today. A department head (*buchō*) of one of Japan's largest petrochemical firms remarked, 'This is a delightful account of the appearance of a "true" *Shingikai*, instead of a bureaucrats' *Shingikai* that is made for bureaucrats and dependent on bureaucrats.'[39]

Since the days of the Electric Power Generating Co. case, MITI no longer nominates businessmen and politicians with vested interests to *Shingikai* or *Kondankai*. Instead, it appoints 'neutral' observers as members, and makes academics (in most cases public university professors) the usual candidates for *Shingikai* chairmanships. In biotechnology policy promotion, all of the subcommittees of the aforementioned Bioindustry Advisory *Iinkai* Committee are headed by Tokyo University professors. The fact that Tokyo University is the most prestigious of all the Japanese universities is significant, since for reasons of credibility, the institution a person represents rather than individual merit as a neutral observer is more important in gaining the AB a suitable audience.[40]

In fact, it is no secret in Japanese politics that a rough guide to assessing the functional importance of *Shingikai* is to check first the fame of the institutions represented by membership and then, the relative ages of the members.[41] Old leaders from well-known universities are often uninterested, out of touch, busy with other things and generally apathetic to the humdrum tasks of analysing data for a report. But these are the very men employed in the most influential *Shingikai*. Young participants from both well-known and less prestigious institutions tend to be more useful at compiling the necessary facts and are found on *Shingikai* involved with the crucial planning stages of a project. Elite, well-respected, elderly scientists from top public universities on AB committees are known as *Gakushiki Keikensha* (men experienced in scholarship). The *Nemawashi* predicts that these are the kind of scientists employed in ABs that handle biotechnology policy issues.

This system of recruiting the 'best' academics for central government ABs is disparaged by critics as 'Boss Politics' (*Bosu Seiji*); the manipulation of top researchers by the civil service to give voices of credibility to steer policy packages around budgetary constraints of the Ministry of Finance (MOF). Academic chairmanships of *Shingikai* and of their subcommittees also are more or less limited to those professors with long-standing association with bureaucratic committees. This suggests that specialized '*zoku*' or 'groups' may be developing in given policy areas among top Japanese academics as already seen with Dietmen experienced in specific fields.[42] Although the ministries they serve argue that these committee members become more knowledgeable of issues and more stable in their positions over time, the reality may well be that these academics are the ones with a predisposition towards legitimizing a ministry's position, and thus are repetitively consulted.

ABs are also important in their contribution to the production of dynamism in the Japanese bureaucratic system. Perhaps their greatest role in this respect is keeping an organization extremely well-informed by providing constant feedback information to promote a response in the face of a changing policy environment. Vital information furnished by an AB, therefore, permits organizational learning and changes of goals in policy matters. Communication between ABs and central government ministries takes place along regularized channels. As information brokers, ABs serve important functions in bridging communication gaps between organizational subunits and for coordinating implementation. *Kondankai* are particularly suited as information 'intermediaries' since they are relatively informal and temporary; more like open meetings or fora for briefing the uninformed on a particular topic rather than 'closed' meetings of restricted participants to thrash out issues thoroughly.

In summary, the *Nemawashi* approach occupies the middle ground of the three component Triple N Synthesis, and focuses on the bargaining behaviour of organizational subunits negotiating a larger, supra-organizational objective. It is based on formal organizations with regularized patterns of human behaviour, legal powers, and procedural rules that persist over time. ABs in the central government are critical of the smooth functioning of this perspective, and give a general consistency to government policies. It is for *Nemawashi* to demonstrate that in the biotechnology case not only does the bureaucracy work as a conglomerate of 'loosely allied organizations', but also that these organizations are the channels through which the bureaucracy rapidly forged a consensus about the importance of biotechnology's development to Japan.

Nawabari Arasoi – the bureaucratic politics approach

The third perspective, that of *Nawabari Arasoi*, is as extreme in its view of 'conflict' as the prime motive in policy matters as *Nebukai* is of 'consensus' and 'agreement'. The Japanese word, '*Nawabari*', literally 'roping-off', here refers to the practice among individual ministries of clearly delineating their spheres of influence. (These territories are also known in Japanese as *shokkan, kankatsu,* or *shokatsu.*) '*Arasoi*' simply denotes a battle or struggle; a word epitomizing the basis of bureaucratic politics – the struggle between actors for power. The act of ministerial administration by 'roping off' specific sectors and the selfish protection of these boundaries is known as *Tatewari Gyōsei* – literally, 'administration by vertical compartmentalization'. Unlike the consensus and conflict view, *Nawabari Arasoi* stresses competition, not bargaining processes, as the mode of interaction among players. But as with most forms of competition, there must be negotiations and bargaining in order for constructive reconcilliation to occur.

In *Nawabari Arasoi*, the players are also different than in the other two perspectives. Since territorial differences are almost entirely between ministries and agencies in the central government, *Nawabari Arasoi* takes 'the-state-as-the-sole-actor' approach to policy-making. In the second component of the Triple N Synthesis, since each ministry perceived itself as a single organization composed of affiliated subunits, there was a sense of 'organizational parochialism' which allowed for a rapid compromise between two or more competing subunits. *Nawabari Arasoi* differs from this since territorial disputes do not originate from intra-organizational struggles among subunits, but inter-organizational conflict between competing ministries. This sometimes produces a situation described as 'a leaderless' or 'blind' policy, since the resultant compromise is often quite removed from the original policy plan of a particular ministry.[43]

The *Nawabari Arasoi* approach for analaysing biotechnology policy is a variant of the classic 'Bureaucratic Politics Model'. It differs primarily in two ways. First, the latter focuses in greater detail on the politics between individual leaders of government departments in determining policy decisions. In Japanese policy, despite a rigid hierarchical organizational system in a bureaucratic unit, a *Nawabari Arasoi* approach suggests an absence of a tendency for seniors to monopolize power in the pursuit of individual political gain (sometimes called baronial discretion). Instead, normative considerations persist, whereby participants act in accordance with organizational guidelines for the good of the ministry. If anything, Japanese bureaucratic competition does not indicate a small number of 'Indian Chiefs' preoccupied with their relative power in government – and thus their own mobility in an organization – but the exact opposite; an acknowledgement by 'Indians' in different ministries of the competing goals and rivalry between their organizations, and the efforts of the 'Chiefs' to resolve the conflicts. Second, the Bureaucratic Politics Model describes regularized action channels with specific rules for approaching controversial policy issues among government branches.[44] This suggests that if in Japanese biotechnology policy, irregularity of behaviour with ministries in conflict is more the norm, ministries would engage in 'unauthorized behaviour' (*ekkengyōgi*) – managing industries outside their jurisdictions – to win conflicts. Furthermore, from initial disputes, persuasion, accommodation, and final negotiation of a compromise would proceed more or less at random.

Competing Japanese ministries tend to have imperialist attitudes when setting priorities for policy. According to *Nawabari Arasoi*, their imperialism concerns aggrandizing three power resources: the procurement of money for purchasing power, the recruitment of educated elites for manpower, and the adoption of new administrative areas for territorial power. Each ministry secures its bid to expand the wealth, subjects and territory of its empire by consolidating independent power bases. The first power base is rooted in Japan's political tradition; the formal authority and responsibility developed by a ministry through the years over a particular policy area. These areas or jurisdictions are initially determined by legal prescriptions, Statutory Laws (*Setchipō*), enacted when a ministry is first founded.[45] As policy practices slowly evolve and the ministry defines its political niche in the bureaucratic system, interpretations of legal orders, conventions and general 'parochial priorities and perceptions' all supplement initial ministerial authority and define organizational responsibility.[46] Second, there is power in what resources a ministry controls. The number and kind of companies traditionally handled by a certain ministry in addition to the specific products administered (whether chemicals, food, manufactured products and so on) are essential in determining power sharing between ministries. Third, there is the quantity and quality of information resources. Bureaucrats specialize in expertise and control over vital information usually through sprawling networks of auxiliary 'think tank' like organizations and their use of ABs.[47] This information is critical to a ministry's task of administrative guidance in Japan, identifying policy options, estimating feasibilities, and eventually evaluating the final action. Fourth, the Japanese government practice of employing distinguished 'experts' on ABs to advise on policy matters opens the door for wielding these committees as instruments promoting the credibility of an unpopular bureaucratic decision. Well-known national specialists could give tremendous competitive advantage to a ministry in its imperialist struggle against another to increase its resources.

Since these struggles are primarily between the different organizations of the civil service, almost all other actors are excluded from the *Nawabari Arasoi* perspective. This restriction of the central players means that policy views on any issue become moulded by a particular ministry's values. As Allison acknowledges, in budgetary conflicts given the situational imperatives of bureaucratic politics, there is no answer to questions such as how much to spend to implement a certain policy.[48] Given a particular policy issue, these value judgements depend on a ministry's perception of that policy's importance; a perception that changes from ministry to ministry.

One result of organizational parochialism and competing ministries is the dynamism produced. Imperialist organizations are vigorous ones.[49] The impetus for revision is not an exogenous force of incremental change due to pressure groups, nor a revolution due to crisis, but a drive from within the organization itself aimed at gaining new strongholds of power. In Japan, the energy and vitality of the bureaucracy is particularly well explained by this approach of the Triple N Synthesis. Dynamism in Japanese ministries is essential to what Pempel termed 'creative conservatism',[50] or the remarkable adaptability shown by the Japanese bureaucracy in response to changing needs of the domestic and international environment. On occasion, this dynamism degenerates into a 'must-do-something' syndrome, so often observed in Japanese biotechnology policy matters in the guise of unnecessary guidelines to delineate jurisdiction boundaries, empty gestures abroad to alleviate short-term international pressures, and so on.

In summary, *Nawabari Arasoi* is concerned with bureaucratic battles for competing goals. These struggles are usually imperialistic in nature with manpower, financial resources and jurisdictional rights at stake. Compromise is the usual outcome of these competitive skirmishes; dynamism and fresh approaches are useful by-products.

Chapter two

Definition of biotechnology

Since this study attempts to analyse Japanese policy towards biotechnology, it is important not only to consider international concepts of this discipline, but also notions of what biotechnology is in Japan. This chapter will attempt to explain the perceptions and expectations of biotechnology as seen by the Japanese. Furthermore, a clear understanding of the biotechnology field will help answer larger questions concerning the structure of Japanese industrial policy, the developmental strategy for bio-technology's implementation in future industries, and the effect of the technology on structuring priorities in a future Japanese economy.

The first use of the word, 'biotechnology' is attributed to two engineers at the University of California who entitled a 1947 article in *Science*, 'Biotechnology: A New Fundamental in the Training of Engineers'.[1] The meaning here is somewhat different from that found in most contemporary journals on science, since for about a quarter of a century, 'biotechnology' was used in the same vein as modern ergonomics; the study of the relationship between workers and their environment.[2] In fact, in 1972, biotechnology is referred to as, 'The branch of technology concerned with the development and exploitation of machines in relation to the various needs of human beings.'[3] Only a year after the Oxford presses produced this volume, the biological sciences were rocked by startling discoveries that changed the entry under 'biotechnology' in subsequent dictionaries of contemporary English. Since there is little agreement about a succinct definition to a discipline so wide in scope as biotechnology, it is best to describe it historically by reference to three distinct generations – even though the word 'biotechnology' as we know it did not become popular until the third generation.[4]

The first generation stretches back for centuries, both in the East and West, where 'artists' skilled in fermentation produced

28

alcohol, leavened bread, cheese, pickled cabbage, *shōyu* (soy sauce), *miso* (bean paste) and *nattō* (fermented soy beans). Similarly, since time immemorial, bacterial action has been exploited in the planting of legumes for enriching weakened soil, in cesspools for decomposing wastes and in making soaps from fats. This first generation of biotechnology is characterized primarily by people developing manufacturing processes by trial and error, and continuing with them because they worked, not because of any formal understanding of the scientific principles involved. Even by the end of the nineteenth century when advances in microbial sciences in the laboratory led to elucidation of some metabolic mechanisms, fermentation at the industrial level continued to operate largely as an art. There was little input from theories of the biological sciences (basic research), nor any contributions from results of experiments specifically designed to increase yields (applied research). Most improvements were practical in nature based on mere chance, or the instincts of the 'artists' concerned (development research).

This second generation can be split roughly into the periods before and after the Second World War. These two periods are loosely related by an underlying continued development in bioprocessing. (Bioprocessing simply refers to the techniques involved in using microorganisms to convert a raw material substrate into a product. These techniques include tank design, speed of feeding nutrients, screening and cultivation of microorganisms, purifying the final product, and so on.) The second generation can also be distinguished from the first by the tendency of 'artists' to recognize the importance of R&D, and to realize the significance of a strong commitment to applied research. Although basic research did play a role in initiating some technical change, the effect was small compared to its importance in the third generation.

The first half of the second generation 'began' during the First World War when interest in fermentation technology was re-kindled due to heavy demands for acetone and glycerol in munitions.[5] The Germans, for example, used fermentation during the war to replace 60 per cent of their pre-war food imports.[6] They grew a yeast *Saccharomyces cerevisiae* for human consumption to supplement sausages and soup. The development in 1914 of the activated sludge process for mineralizing organic waste – still employed worldwide today to treat sewage – was also due to applied research during this period.[7]

The contemporary strength of Japan in fermentation is largely derived from the active academic pursuit of bioprocessing and the

Biotechnology in Japan

efforts of the *Nihon Nōgei Kagakukai*, the Japan Agricultural Chemistry Society. Since 1927, members of this society have not only conducted research in agricultural chemistry, but have experimented in a wider range of topics. Japanese style agricultural chemistry is explained in a guide for undergraduates admitted to Tokyo University's Agriculture Department as follows:

> It is difficult to explain what *Nōgei Kagaku* (Agricultural Chemistry) is in one sentence. If one looks at the papers written by specialists in the Japan Agriculture Chemistry Society and other related academic associations, you'll be surprised at the wide range of research taking place. In short, it is a field which exlains, in chemical terms, phenomena associated with animals, plants and microbes, and uses this information for the improvement of human life.[8]

The guide continues that there is no real equivalent of this field in the US or Europe, since it is so interdisciplinary in nature. In fact, 'Agricultural Chemistry Society' is vague English for an organization whose members study chemistry, biochemistry, microbiology, and genetics as they relate to agricultural production, the food industry, medical pharmaceuticals, energy, the environment, and so on. The similarities to third generation biotechnology (described later) are considerable. Japanese scientists from this society were responsible prior to the Second World War for the Aji-no-moto (origin of taste) company's emergence as the world's first manufacturing distributor of monosodium glutamate (MSG), the most economically important of all the amino acids. It was also these scientists, who in the latter half of the second generation, responded to Japan's rapid development in the fermentation industry.[9]

Penicillin galvanized Japanese fermentation efforts during the US Occupation in 1946–47, when strains of the mould penicillium were introduced from the US. Over seventy companies rose to the challenge of producing penicillin; only four survived. The fierce competition that ensued brought forth many innovations both at the microbial and fermenter levels of technology. This ultimately led Dr Kinoshita and his collaborators at Kyōwa Hakkō Company in 1956 to develop the first metabolic regulating fermenting technique for micro-organisms enabling production of the amino acid, glutamate.

Subsequent improvements in fermentation technology enabled Tanabe Pharmaceuticals to discover another world first in 1969 – a technique for manufacturing amino acids from the immobilized enzyme, amino acylase.[10] These advances in turn paved the way for Japan to dominate today's amino acid market which accounts for two-thirds of world production, to lead the world as a supplier of antibiotics satisfying 60 per cent of world demand, and to supply over 40 per cent of the world output of high fructose syrup.[11] By 1970, Japan had a ten-year lead in fermentation over the rest of the world using traditional 'artistry' in the fermented foods industry and the learned skills of second generation bioprocessing.

Although the third generation of contemporary biotechnology is acknowledged as beginning in the early 1970s, the technology is rooted in the significance of the discovery of DNA (deoxyribonucleic acid) as the fundamental component of genetic material by the Nobel Prize laureates, J. D. Watson and F. H. C. Crick in 1953.[12] The recent excitement over biotechnology, however, is due primarily to the development of a technique that allows DNA to be altered in the research laboratory. This technique, 'perfected' in 1973 by Stanley Cohen and Herbert Boyer from the University of California in San Francisco, is called recombinant DNA – also known as genetic engineering, gene splicing, gene cloning or simply as rDNA. The procedure involves changing the metabolism of a microbe at the genetic level to yield a 'new', slightly altered metabolism with novel capabilities such as the ability to metabolize harmful or toxic wastes, to function at extreme temperatures or to produce enormous amounts of a specified protein. This exploitation of the microbial chemical factory has been aided by a long evolutionary process whereby microorganisms have adapted to bizarre conditions. Prospects for industry are obvious. In fact, most of the current R&D in rDNA centres on methods for genetically manipulating microbes to render their metabolisms more 'suitable' for industrial processes.

Another key advance in third generation biotechnology – only slightly dwarfed by rDNA – was the discovery in 1975 of cell-fusion technology. These techniques were developed by another Nobel Prize laureate, Caesar Milstein, and involved the creation of monoclonal antibodies that could be directed towards one type of material foreign to the body (antigens). Antibodies are proteins capable of several important immunological functions. They both recognize and bind antigens, leaving the foreign material susceptible to destruction by both immunological cells (T-cells and macrophages) and the body's complement system – a group of twelve enzymes involved in degrading foreign antigens. Monoclonal antibodies are special since they are extemely specific and sensitive, binding only to one type of cell which enables them, in theory, to distinguish between normal and cancer cells. If monoclonal antibodies could be targeted towards a tumour, they could be used to guide protein toxins (diphtheria, ricin, botulin) directly to the growth, thus destroying it locally. Monoclonal antibodies have recently proven to be quite effective as early diagnostic tools for monitoring the onset of various diseases, such as the Enzyme-linked Immunadsorbent Assay (ELISA) test for confirming the presence of the AIDS virus, HIV, in sera. Briefly, in an ELISA assay or test, a patient's blood is added to a test tube filled with AIDS-infected cells. Certain antibodies in the patient's blood that have naturally developed over time to the HIV virus bind to these infected cells. Laboratory-produced monoclonals specific for the patient's antibodies are then added, and these fix themselves firmly on top of the patient's antibodies. Since these monoclonals have an enzyme attached to them, a chemical can be added to the solution that reacts with this enzyme. The chemical reaction that ensues produces a readily visualized colour directly over the spot where the monoclonal, and ultimately, the infected cell sits. This appearance of colour suggests that a patient has antibodies in his blood specific for the AIDS virus and thus is a positive test for AIDS.

Third generation biotechnology, one of the few technologies that is rooted in basic science, is derived from the first principles of the biological sciences. This is quite distinct from the development and applied kinds of research associated with the first and second generations respectively. Biotechnology has thus matured from an 'art' into a rigorous discipline requiring a clear understanding of fundamental processes of the biological and physical sciences. Many of the bioprocessing innovations of the fermentation industry from the second generation, however, are still very central to biotechnology in the third generation; technologies such as immobilized enzyme techniques, large-volume cell cultivation, microbial screening and metabolic suppression. Third generation biotechnology is different from the other two because of the introduction of novel 'inputs' such as rDNA technology, monoclonal antibody production, automatic DNA synthesizing processes and so on. Although there is an acrimonious debate about whether or not biotechnology is a science or a technology, it is considered in this book to be a technology in the sense of 'use-oriented knowledge' (see Chapter Three); an assimilation of many experimental techniques and processes.

Recent advances in equipment and a shortened doubling rate for scientific knowledge have also widened the potential growth of third generation biotechnology. The automation of bioprocesses with on-line computers, continuous sensor devices for selection and screening of fermentation broth, and the development of research which integrates biology, electronics and computer technology are central to pioneering work such as biocomputers, capable of artificial intelligence, bionic limbs and organs for the handicapped, biochips that function with electric circuits built on the molecular level and so on. It is because of the potential for new discoveries that the popular press has described second generation biotechnology with its bioprocessing as 'old' and third generation biotechnology with its more powerful techniques as 'new'. In this light, an operational definition for the international perception of the new biotechnology is aptly described by the US Congress Office of Technology Assessment (OTA) as, 'any technique that uses living organisms (or parts of organisms) to make or modify products, to improve plants or animals, or to develop micro-organisms for specific uses.[13]

In Japan, the new biotechnology is readily distinguished from the 'old' and noted for its diversity. The foreword to a recent MITI publication on biotechnology policy explains:

Biotechnology is a technology that has been used primarily in the industrial fermentation field with things such as organic chemicals and brewing products. At present, however, biotechnology is rationally fusing the functions of microorganisms for biological synthesis, metabolism of substances, and so on, with chemical manufacturing ... More recently, due to advances in molecular biology, molecular genetics and so on, recombinant DNA technology has become established. That is to say a new biotechnology has rapidly developed with its *potential* being extremely broad.[14]

Although written in the main in katakana, and thus retaining its English pronunciation, the English 'biotechnology' causes some problems when written in Kanji (Chinese characters). The most familiar Kanji translation, often appearing parenthetically after the katakana form, is *seimei kōgaku* literally 'the engineering of life'. More direct equivalents of the English are *seibutsu-ryō-gijutsu*, 'technology-using-biological organisms' or *seibutsu-kōgaku-gijutsu*, 'biology-engineering-technology'. It is probably because of the versatility of Kanji that an even larger proliferation of words describing biotechnology has appeared in government documents, industrial reports as well as in the lay press; words such as *seibutsu-kei kagaku*, 'biology-like science', *seimei-gijutsu*, 'life technology', or *seimei-kagaku*, 'life sciences'.

In the various ministries promoting biotechnology, the technology is seldom referred to by the word 'biotechnology', each government office having a preferred substitute. MITI, for example, has created the word bioindustry similar to the Kanji equivalent, *seibutsu kōgyō*, 'biological industry'. This definition gives a stress to MITI's dedication of developing biotechnology into a fully fledged industrial endeavour. The Ministry of Education (MOE) uses the term 'bioscience' since it views biotechnology as merely a conglomerate of different technologies such as bioprocessing, genetic engineering and hybridoma technology, all rooted in the same basic biological sciences of molecular biology, biology, genetics and microbiology. First used in 1971 before the discovery of rDNA, the term *raifu saiensu* or life science has been made popular by the Science and Technology Agency (STA).[15] The Ministry of Agriculture, Forestries and Fisheries (MAFF) has created a new permutation of *baioteku* by leaving out the 'o' to give *baiteku* which gives a word similar in spelling to the Japanese for high tech, *haiteku*.

Biotechnololology in the popular press

In the popular literature the definition of biotechnology has been very colourful. The sensationalism in the mass media about biotechnology's influence on so many facets of life has left questions about the extent of its impact. The Bioindustry Development Centre (BIDEC), a private think tank of MITI, predicted in a 1985 report, that biotechnology-based industry may account for about 10 per cent of the Japanese gross national product – about ¥127 trillion – in the year 2000.[16] Metaphors created by the Japanese press described the optimism towards biotechnology as bioboom (baiobuumu) and biofever. Japan has not been the only country ensnared by 'biotechnology fever'. Both the US and Europe exhibited similar symptoms of public excitement in the early 1980s with rapid rises in the stock prices of new biotechnology firms (NBFs), and a lucrative flow of investment from established chemical and pharmaceutical companies.[17]

It is probably safe to assume that during the 'baiobuumu' period of the early 1980s, the Japanese mass media carved their own definitions of biotechnology for the public. As these definitions became increasingly all encompassing, infringing on all aspects of the biological sciences, the suffix of 'technology' was dropped, and only the use of 'bio' (Baio) became popular. In modern Japanese, 'bio' is able to appear alone unlike in English (except in American slang where 'bio' means a resumé or biography of a person), and highlights Japanese as an example of a living language that adopted the Indo-European prefix and retained its Greek meaning, 'of living things'. Books entitled *Baio no Sekai* (*The World of Bio*), *Baio no Chosen* (*The Challenge of Bio*) or, '*Baio Gyōsei o Ōu*' (*Following Bio Administration*) illustrate popular uses.

Although convenient for capturing the broad implications of biotechnology in Japan, where everything from medical equipment to electronic limbs is considered part of the discipline, the word 'baio' creates much confusion. Biotechnology, described in this way, draws boundaries too broad for serious discussion. In any given project, for all relevant aspects of the field to be fully addressed, 'bio' demands a team of specialists too large for much practical R&D success. Too general a use of 'bio' tends also to imply that the technology is a panacea; a miracle cure for Japan's poor endowment of natural resources and a simple prophylactic for future economic problems.

Biotechnology in science

In the science field the real litmus test of a biotechnology definition becomes the actions of the scientific and industrial researchers involved in the discipline. There is extensive debate about whether this action is either a response to both a definite government policy and a popular consensus, or whether it is itself the initiating force of the government's policy and consensus. This will be considered more fully later. What is important here is that the three objects of most funding in Japanese biotechnology, rDNA, cell fusion and bioreactors, provide a *de facto* definition of the technology. Saitō Hyūga surmises:

> The meaning that the word 'biotechnology' has in Japan is a little different from the meaning that it has in the United States and Europe. In Japan its meaning is in terms of recombinant DNA (rDNA), cell fusion and bioreactors. However, in countries other than Japan, the word 'biotechnology' has stronger connotations of bioengineering, cultivation technology, and cultivation simulation in the main.[18]

But as Imada notes, rDNA, cell fusion, and he adds large-scale tissue culturing, are components of the new Japanese biotechnology. Only bioreactors refers to second generation bioprocessing techniques of the 'old biotechnology'.[19] As research on the 'leading edge' continues, biotechnology is appropriately redefined in order for it to keep pace with scientific advances. The assumption throughout this book is that, if no reference to the distinction between 'new' and 'old' is made, the biotechnology referred to is that of the new third generation.

Biotechnology has grown quite rapidly in the past fifteen years spurred by advances in the supporting basic sciences, especially microbiology, cell biology and genetics. Genetic-engineering techniques can now be applied to plants, amino acids and DNA sequences are quickly elucidated, 'gene machines' are employed for synthesizing genetic material, and monoclonal antibodies are being produced routinely both in the laboratory and on a larger scale in industry. What does Japan expect the 'new' biotechnology to contribute to her future?

Expectations of biotechnology

A 1983 document by MITI, appropriately dubbed Biovision, explains that the government's expectation for biotechnology is to increase the specialization and modernization of the private sector in order to satisfy new market demands:

> The nation's needs are becoming more diversified with rising incomes and changes in lifestyles, and the private sector, in order to handle this, is promoting specialization and a move from the production of large quantities of a small range of items to the manufacturing of small quantities of a large range of items. Biotechnology will contribute much as a technology able to respond flexibly to these diverse demands (of the nation).[20]

The strength of biotechnology is its ability to be applied as a basic technology in a wide variety of industrial sectors. There are certain fields for which biotechnology is thought particularly advantageous: medicine, petrochemistry, pharmaceuticals, agricultural chemicals, synthetic fibres, synthetic gums, chemicals, textiles, food products, energy sources and protection of the environment.

Modernizing industrial structure

Biotechnology enables specialization of the Japanese industrial structure in two areas: the supply of new, high value-added products and the provision of more efficient resources. Already rDNA technology has made available previously unobtainable high value-added products such as interferon, insulin, and human growth hormone. Other direct applications of biotechnology include the mass production of fine chemicals, of new vaccines, and the like.

The integration of biology and electronics, or bioelectronics, promises to increase the efficiency of industrial processes by saving energy, increasing output, and producing less waste. Biosensors, for example, could prove economical in process management in the food and chemical industries. Continuously operational bioreactors could replace outdated fermentation tanks in ethanol, amino acids, and organic acids production. Biomass, defined by Higgins as 'the overall concept of producing high-grade fuels and specialty chemicals from purpose-grown plants or biological wastes', offers renewable fuel alternatives to the exhaustible fossil oils; a welcome prospect to Japan, infamous for its lack of adequate natural energy resources.[21]

Elevation of medical welfare

Although Japanese medical technology has advanced very rapidly, the persistence of terminal illnesses like cancer together with rising health costs and the increase in the ages of patients have resulted in a 'large societal problem'.[22] The development of inexpensive pharmaceuticals, diagnostic products and sophisticated medical equipment could alleviate many difficulties, and has made the potential of biotechnology very attractive.

Biotechnology, however, is still a youthful industry with few products in relation to the huge investments in R&D. In medicine, the fruits of research are just beginning to reach the market. The growth of biotechnology in medicine, as with most other areas, has been largely determined by simple supply-and-demand economics. Medical research for the leading terminal illnesses in the industrialized nations – cancer, cardiovascular disorders and diabetes mellitus – has received large allocations of funds; research for tropical disorders has been routinely ignored. The US Institute of Medicine estimates that in the US alone, cancer research increases at a rate of $209 per patient per year, heart disease at $8 per patient and the parasitic diseases of schistosomiasis and malaria at 4.5 cents and 2 cents per patient per year.[23]

Diabetes, heart disease and cancer

It comes as no surpise then, that the first drug genetically engineered and licensed for human use was insulin, taken by 15 million people in the developed world as a therapeutic against diabetes. Eli Lilly & Co. together with Genentech, distributed this product, Humulin, in 1982, four years after successfully cloning the gene in the laboratory. It was subsequently marketed in Japan. The rDNA preparation is less expensive and, more importantly, provides less of an opportunity for an allergic reaction than the bovine insulin employed previously.

The human interferon gene has also been cloned and the rDNA interferon produced. Before cloning, interferon was obtained by direct extraction from human cells which meant that only one gram of interferon could be obtained from over 90,000 donors at an estimated price of $50 million per gram.[24] By 1983, Japanese companies were actively competing in the interferon business and reducing the price of one gram by 90 per cent. Interferon is an antiviral agent which seems to benefit patients with cancers of the skin, bone, breast and blood. Most of the enthusiasm surrounding the use of interferon in the late 1970s and early 1980s was due to its potential as an anticancer agent. On the whole, most of the varieties of interferon have proven less effective in treating cancer than once thought. Other drugs such as the lymphokines, especially interleukin-2, are receiving more attention as possible cancer therapeutics. These drugs enhance the proliferation and activity of natural killer T-cells which seem to recognize tumour cells as foreign and destroy them.

The biotechnological industry can also be applied to the treatment of disease. A recurrent cause of cardiac arrests often involves the formation of blood clots that stop the flow of blood, and hence oxygen, to heart muscle cells. Biotechnology has enabled the production of large quantities of enzymes capable of dissolving these clots. Drugs incorporating these enzymes have potentially huge markets in Japan where circulatory problems, which accompany old age, are sure to become more prevalent in Japan's ageing society. One of the most successful of the circulatory therapeutics, urokinase, generates annual sales in Japan totalling £100 million.[25] Using biotechnological techniques, urokinase can be derived from a microorganism inserted with the human kidney cell gene; previously Japan extracted the enzyme from urine imported from South East Asia.[26]

Although urokinase is excellent in decreasing the ability of the blood to clot, it does so throughout the whole body thus allowing a risk of internal bleeding without the defence of the body's clotting mechanism. Other drugs, such as tissue plasminogen activators (tpa), are more localized and precise agents for dissolving blood clots. Tpa produces an enzyme plasmin that catalyses the reaction for unclotting blood. Tpa functions by attaching itself to clots and by riding 'piggy-backed' around the body until the clots disappear. In Japan, many companies have been active in joint international cooperation projects for tpa production.[27]

Infectious diseases – vaccines, antibiotics

Vaccines can be developed less expensively and far more safely using biotechnology rather than conventional techniques. In 1967 when the World Health Organization (WHO) finally obliterated smallpox from the world, the conventional method of inoculating a weakened form of the virus was used which incited the production of antibodies against it.[28] Even though this vaccine contained weakened viral DNA, there was always a risk that the DNA would be incorporated into healthy cells, bringing about smallpox. Genetically engineered vaccines, on the other hand, no longer have this risk. In the manufacturing process, a specific antigenic part of the viral or bacterial proteinaceous coat is identified, the gene for this area is cloned and the pure protein product is injected as the vaccine. Since the coat is foreign material, the antibodies will respond, but the absence of any viral/bacterial DNA rules out possible infection. Vaccines against hepatitis B, polio, yellow fever, rabies and rubella are now all manufactured in this way.

It is unfortunate, however, that R&D for human vaccines against other common worldwide diseases such as influenza, herpes simplex, mumps, measles, common cold and shingles are progressing more slowly. One of the problems with the common cold, for example, is that instead of just one virus being involved, there are over 200 different types, each recognized by a different antibody. Other impediments to development include the stringent requirements in most countries for testing human vaccines which have prompted private firms to turn increasingly towards animal vaccine production for profit (see later under livestock).

One exception to this is the continued improvement of the vaccine against hepatitis B, a viral infection of the liver. In Japan, hepatitis B has quite a high incidence especially in Kyushu, and although a rDNA vaccine has been available since 1981, it is expensive. The government and the private sector are actively pursuing R&D for a more efficient and economical vaccine. In 1986, for instance, the STA allocated ¥1.1 billion for such a vaccine.[29]

In 1975, the United Nations Development Programme, the World Bank and WHO set up a Special Programme for Research and Training in Tropical Diseases.[30] They cited six diseases, most prevalent in the Third World, for which vaccines or effective treatments must be sought; malaria, leprosy, filariasis, leishmaniasis, schistosomiasis and trypanosomiasis. Some successful work is being done with malaria, leprosy and leishmaniasis, but again, there is no economic incentive for companies to fund projects with a large risk of failure and no lucrative markets.

Although over 5,000 antibiotics have been identified, only 100 or so are available for human use.[31] The majority are produced from actinomycetes, similar to moulds in appearance. There are four main classes, penicillins, tetracyclines, cephalosporins and erythromycins, with worldwide sales near $4 billion per year.[32] In the antibiotics industry, there is necessarily a constant attempt to produce new antibiotics or related analogues because of 'learned' bacterial resistance. Bacteria have not only become resistant to the earlier forms of penicillin, for example, but have also completely nullified the drug's use in some instances by producing an enzyme, penicillinase, that attacks the antibiotic even before it can do its job. Resistance to the newer antibiotics is a particular threat to hospitals which provide ideal settings for the contagious spread of iatrogenic infections. Recently developed antibiotics are thus given sparingly with the hope of evading resistance to fast mutating bacteria. Most of the antibiotic analogues today are still chemically formed from a handful of base products; 6-amino-penicillanic acid (6-APA), for instance, is the base from which 40,000 semi-synthetic penicillins are prepared.[33]

Medical equipment

An exciting frontier in medicine involves the infusion of advances in electronics, engineering and computer science with biology. Sophisticated medical equipment such as nuclear medical resonance (NMR) and advanced medical information systems such as MYCIN (a computer database network in California) are already widely used for diagnosing and treating diseases. Computer databases are particularly useful in staying abreast of the myriad new drugs approved worldwide annually. There is expected to be a proliferation of medical biosensors which would regulate the quantity of biochemicals, enzymes and antibodies in the blood. Glucose biosensors, for example, implanted inside the body would precisely regulate the blood sugar level for sufferers of diabetes. This could prove beneficial to chronic sufferers of diabetes since it is believed that the lack of fine control in glucose levels, over the long term, is responsible for secondary, life-threatening effects of this disease.[34] There are also real possibilities of producing artificial limbs and enabling the permanent implantation of a wide array of artificial organs.

Food

By the year 2000, the world population is expected to exceed six billion, but there is a projection that cereal production will not be able to accommodate this population.[35] In Japan, the supply of food produced domestically has been decreasing each year. MITI's Biovision report quotes the most recent figures as Japan having a 73 per cent overall capacity to supply its own food, 53 per cent capacity of supplying the required calorific intake of the nation, and only 30 per cent capacity of supplying the needed cereals: 'In the future, there will be unavoidably a heavy dependence on countries abroad for food.'[36] Biotechnology is expected not only to provide large quantities of domestically produced food at low costs, but also a diversity of food varieties to fit changes in tastes.

Agriculture – nitrogen fixation

Inputs for nitrogenous fertilizer constitute a large part of the total cost of a market crop. Reducing crop dependence on fertilizer is, therefore, a fundamental goal of biotechnological R&D in agriculture. Farmers have known, through trial and error, that rotating legumes such as peas, beans, clover, peanuts and soybeans, with other crops, restores the nitrogen content of the soil. This is due to *Rhizobium* bacteria which fix atmospheric nitrogen in the root hairs of legumes. The bacteria contain a dozen or more nif genes (*ni*trogen *f*ixation), and produce the enzyme nitrogenase, which catalyses a reaction forming ammonia from atmospheric nitrogen. Most major cereal crops, such as rice, wheat, corn and barley cannot fix atmospheric nitrogen since *Rhizobium* bacteria are unable to live in their root hairs. If fertilizer is to be made truly obsolete, either *Rhizobium* bacteria or cereal crops must be modified artificially to be compatible with each other, or the plant itself must be genetically altered to fix its own nitrogen.

Creating a superplant

Many biotechnological applications in agriculture centre around the cultivation of plants with increased yields and with tolerance to adverse conditions of aridity, salinity, temperature and the onslaught of disease. Plants could be grown for food in currently hostile areas such as marshes, areas in low rainfall or very hot or cold climates. Natural selection has already given rise to trees like the river redgum of Australia which has adapted to very salty water. Biotechnology has strengthened this selection by providing the technology for the cloning of genetic replicas of the most well-adapted river redgum tree. The Unilever Company has exploited this method of cloning the 'fittest' tree in its palm oil project in Malaysia, producing a thousand or more nearly 'identical' copies of a single high-yielding palm. Annual palm oil sales have soared to over $2.5 billion per year since the late 1970s.[37]

Another technique for transferring genetic information in plants is by cell fusion – joining of two cells stripped of their protective cell walls. Natural recombination processes (like the joining of a human egg with sperm) recombine genetic material in this way. The hybrids created can be selected for useful characteristics, although the technique as a whole, lacks the precision of genetic engineering.

Research has also revealed strains of soil bacteria that secrete toxins which kill pests such as nematodes. Associating these soil bacteria with plants or by inserting their genes into a plant, may alleviate the need for pesticides and herbicides. The Calgene company has been working on cloning the gene(s) responsible for immunity to herbicides and inserting them into soybeans. In this way, whole fields sprayed with herbicides would be effectively 'weeded' thus reducing the frequency and cost of tillage to eradicate weeds. Crop production would also increase due to reduced competition among plants.

Genes concerned with a plant's ability to resist climatic stresses, osm genes, have also been isolated. There is the possibility that inserting the gene for the amino acid lysine into cereal crops may give a more balanced blend of amino acids in our vegetable diet.

Single-cell protein and livestock

In addition to creating more nutritious plants, biotechnology can be applied in growing and harvesting microbial mass which itself is a food product rich in protein. After World Wars I and II, brewer's yeast and *Candida* yeast species were used to replenish diminished food resources. In the 1960s, several firms, such as British Petroleum, invested heavily in microorganisms for human consumption. This food is termed single-cell protein (SCP) to distinguish it from protein produced from higher, multicellular organisms. The central operation in SCP production involves fermentation technology in which the main object is the optimal conversion of substrate into microbial mass. There are two important attractions of SCP.[38] The microorganisms can be grown quickly because of their fast doubling rate, therefore assuring rapid availability of food, and also, the substrate can be varied depending on the microbe chosen for cultivation. Despite these advantages, the SCP production of the 1960s failed to attract a suitable market. One of the crucial factors for SCP's commercial failure, as a human food, was its novelty. In Japan, the government and consumers alike were very sceptical about a product made entirely of microbial mass, microorganisms that had no prior usage or accepted occurrence in any food. The aesthetic factors associated with eating unscented, lightly coloured, powdery 'bugs' dissuaded many opinions. Having had a brief 'flirtation' with SCP in the late 1970s, most Japanese companies have left it to the Europeans in the current wave of new biotechnology activity.

Livestock improvement has become an exact science especially with the advent of artificial insemination technology and industrial production of hormones. These tools are regularly employed to raise animals that give more meat or more milk or both. One hormone, bovine growth hormone, for example, can increase a cow's milk production by 40 per cent and cause 10–15 per cent weight gain.[39] In recent years, hormones have come under attack as leaving carcinogens in the beef from treated animals.[40] This has prompted research investigating the use of alternatives, such as monoclonal antibodies, to attack and destroy animal body fat resulting in a cow with leaner meat. Eventually, these monoclonal antibodies will be available for people wanting to lose weight.

The economics of production 'coerced' the first vaccine produced by rDNA technology to be one for animals. The vaccine manufactured in 1982, was against scours, a bacterial diarrhoeal disease affecting new born calves and piglets. Most of this product is synthesized by the Dutch company, Akio. The United Nations Industrial Development Organisation (UNIDO) gives high priority to the development of genetically engineered vaccines for livestock. Large gains are being made against foot-and-mouth disease, rabies, African horse sickness and blue tongue. Animal interferons are also being researched with the bovine interferon gene having already been isolated.[41]

Energy-biomass

The sudden rise in oil prices in the 1970s, precipitated by the Organisation of Petroleum Exporting Countries (OPEC), forced the entire world to consider alternative forms of energy. In the 1970s, Japan in particular was vulnerable to shifting world oil prices and turned to biotechnology, which offers a variety of basic natural energy conversions using microorganisms, as both a provider of alternative energy sources and a supplier of a reusable energy store. The most promising of the technologies is biomass. In Japanese Kanyi notation, biomass is often written as energy derived from 'the mass of plants' or from 'the present mass of plants'. Simply stated, it is the harnessing of solar energy by the photosynthetic systems of plants.

The sun is the ultimate source of 99.4 per cent of all non-nuclear energy on earth; all the coal, natural gas and oil deposits are linked in an energetic chain leading back to the sun. Photosynthesis is incredibly inefficient at capturing the total available solar energy. As much as 50 per cent of energy is unavailable simply because the photosynthetic activities of plants are limited to light waves between 400 and 700 nm in length. The amount of total solar energy finally entrapped in the leaf at the end of photosynthesis is at best 3 per cent, with most plants averaging efficiency rates of less than 1 per cent. It has been calculated that if the earth had only 0.1 per cent of its surface covered by plants and harnessed solar energy at an efficiency rate of 10 per cent, all current world energy needs would be satisfied after fermentation of these plants to fuel alcohol.[42] Increasing photosynthetic efficiency is a very difficult task, not thought to be generally feasible for many years to come. Short-term approaches concentrate on identifying plants which do exhibit high rates of photosynthetic efficiency, and using them as substrates for the production of competitive fuels.

Today's ideal energy plants, that is, crops which are readily fermented, achieve 3 per cent photosynthetic efficiency requirements and have relatively high energy output to input ratios. Invariably these plants are food crops like sugar beet, barley, and maize. As a rule, the increased use of food crops for energy pushes world food prices higher. Kenya, for example, with its rapid population growth suffered from a drastic increase in food imports in 1982 and was forced to withdraw its support for a domestic fuel alcohol programme in which food crops were used as microbial substrates to ferment alcohol.[43] Brazil is the only country where real success has been achieved in fuel fermentation – four billion litres of gasohol (a mixture of gasoline and alcohol) are produced annually.[44]

Safety and regulations

The historical development of rDNA guidelines in Japan is treated in detail in Chapter Four. The Japanese guidelines for biotechnology are based on the 1976 American National Institute of Health (NIH) guidelines which, in turn, were derived from discussions fostered at two international conferences: the International Conference of New Hampshire in June 1973, and the conference at the Asilomar Centre, California in February 1975. These conferences were largely the result of acrimonious debates in the 1970s, especially in the US, over the safety of rDNA technology. The main criticism of the new biotechnological tool rested on the fear of creating novel pathogenic bacteria leading to uncontrollable epidemics. Two important pieces of evidence supported this claim: *Escherichia coli (E. coli)*, a bacterium traditionally found in the human intestine, was the central vehicle in genetic engineering, and plasmids that are used to 'ferry' foreign genes to *E. coli* are capable of infecting a wide range of human cell types.

Since the 1970s, partially due to the NIH's lead, world attitudes towards biotechnology have relaxed appreciably, with many of the more stringent rules abolished. A decade of research has revealed that the genetic manipulation of pathogenic organisms must incorporate certain preliminary steps which cannot be quietly dropped from research protocols, that all laboratories have organisms in air-tight chambers with little chance of their escape, and finally, that *E. coli* K-12 and *E. coli* X1776 strains, which are generally used in experiments, are feeble and readily perish in the human intestine.

Still concern persists. Recent test cases in the US for the right to release microbes into the environment in experiments have met with bitter resistance. Researchers at the University of Berkeley in California, for example, applying for the right to use *Pseudomonas syringae* in the open field ran into stern opposition from environmentalists. These slightly mutated bacteria would substitute for natural (wild type) variants and protect plant leaves from freezing during cool weather. The environmentalists, however, who point to the 'inherent' risks of the experiment, and the possible catastrophic effects it could have on the environment.[45] Precedents set by the US are always closely watched in Japan. As Uchida Hisao of Japan's Committee on Genetic Manipulation who advised in creating the Japanese guidelines stresses, 'Introduction of a system from the US is always a prudent choice to secure social acceptance and support to start something new in Japan.'[46] In recent years, Japan has allocated large amounts of money to biotechnology R&D, and has 'targeted' it as the basis of a new strategic industry.

Chapter three

Building a high tech society in Japan

Since the publication of MITI's *1980 Vision*, the *Nebukai* coalition has developed a national agreement on the importance of high technology to the future economic prosperity of Japan. Based on the intense research and development (R&D) activity in high technology industries in recent years, it is often unclear to what extent the consensus about the unlimited potential of high tech is due to sensible forecasts of merit, or to its excessive attention in the public eye.

The *1980 Vision* demanded the fostering of growth in electronics and information technology, new materials, and biotechnology. Shortly after publication, the mass media, business reports, books, and government documents embraced these three fields almost unanimously throughout Japan as areas warranting high priority development. In time, the promotion of biotechnology, electronics (information technology), and new materials as potential driving forces of a future Japanese economy gained much credibility, and have become generally accepted. So much so that most Japanese regard the growth of such sectors as a necessary cornerstone of national R&D strategy:

> The Japanese economy has grown and expanded to occupy one tenth of the current free world economy, but the main force to develop and maintain this economy is represented by electronics, new materials and biotechnology, the so-called high technologies. It is because of this, that research and development in high technology and the promotion and nurturing of high technology industries is strongly called for as a basis for supporting the Japanese economy.[1]

In promoting high technology-based industries, the *Nebukai* coalition has also recognized that no longer can the private sector rely on imported technology. In biotechnology, especially, it must drive to be a fundamental innovator in science and technology through a strong commitment to basic research.[2] In stressing the importance of high technology in revitalizing the economy, the *Nebukai* coalition has recognized a chain of three important promotions: (1) creativity is a vital factor in fundamental research; (2) fundamental research leads to growth in science and technology and (3) developments in both pure science and applied technology provide bases for high technology industries.

It is due to this emphasis on creative, basic research that the coalition is treating the promotion of biotechnology differently than it did steel, cars and consumer electronics in their formative years. The demand for creativity in high technology necessitates not only close government–industry relations as in the past, but also the promotion of long-term, high risk, large-scale basic research projects in national research institutes and universities. This has led to what the Japanese call, '*San-Gaku-Kan*', a large-scale collaboration R&D arrangement of industry (*san*), universities (*gaku*) and government (*kan*).

Why does Japan want to promote biotechnology?

A shift to knowledge-intensive industries

In 1973, the Ministry of International Trade and Industry (MITI) published *MITI's Vision for the 1970s* that stressed as its main theme, 'the transfer (from heavy industrialization) to an industrial structure of knowledge-intensive technology'.[3] This report was the mixed result of several influences.

First, at the time of the MITI publication, there was a worldwide tendency to re-evaluate the importance of technical innovations. The expectations of technological development looked very appealing to a world suffering from slow growth and economic stagnation. In Japan, the mysticism of technology as a potential 'black box' of economic growth was captured in government documents, the media, and business reports. The metaphors labelling the period were colourful – 'a techological revolution', 'an aim for a second renaissance', 'a new wave of industrial society'; high technology was even described as, 'Today's Messiah'.[4] The Japanese were sensitive to the investment trends of other nations, and respected the latent power of technology as a catalyst for economic activity.

Second, during the era of high economic growth in Japan, the economy was based on capital-intensive industries, what is often referred to as a bulk or mass production-economy (*butsuryokei keizai*). These industries, namely iron and steel, cars and shipbuilding, were very expensive in capital expenditure and consumption of scarce resources. New technology in knowledge-intensive based industries was expected to lower capital costs and lead to a curtailment of resource consumption. This was anticipated to mean stronger long-term competitive power for Japanese firms internationally.

Third, the last two decades have marked, for many economic historians, a new global *élan* – an unprecedented era in the growth of science and technology. Whether due to changing market needs, competitive pressures, or the rapid accumulation of knowledge, the world has been on the verge of remarkable technological breakthroughs equal in significance, some argue, to the agricultural revolution of early man, or the eighteenth century industrial revolution in Great Britain.[5] According to this thesis, the shift to knowledge-intensive based industries in the early 1970s was, therefore, an inevitable result of rapid advances in new technology.

The word 'high technology' has been described as 'the heightening of recognition of technology in the knowledge-intensification of industry'.[6] Typical high tech fields are concerned with creating state-of-the-art, high value-added products, such as semiconductors and integrated circuitry, using sophisticated technology and well-trained science personnel. Japan, with the second-largest GNP in the OECD has the capital and engineers with the required levels of education to compete with the US, where 90 per cent of the fastest-expanding industries are in high tech.[7] But, knowledge-intensive industries, by their very nature, call for a strong basis in fundamental research; an area where Japan has traditionally been weak.

In Japan, as in the West, there has been a stablization of growth and decline in the number of people in secondary industries.[8] But in Japan this decline in heavy industry has been accompanied by a new expansion and diversification of the high tech sector. High tech R&D demands long gestation periods since it is only knowledge gained through long-term fundamental research that enables future industries to prosper. Although imported technology has played an important role in the postwar growth, there is a consensus that future economic prosperity can only be gained through scientific inquiry carried out in Japan:

> As a nation without natural resources but with tremendous industrial strength, Japan must now, especially, promote R&D rich in more creativity, and must be responsive to, not copy from, world discoveries. She needs not to produce simple

technology, but established science capable of being sold abroad in integrated systems; Japan needs to aim to be a *gijutsu rikkoku* – a nation built on technology.[9]

High technology's lack of dependence on foreign resources

After the oil crisis of 1973/74, Japan adopted a crisis management strategy (*kiki kanri*) that advocated ample preparation for a crisis such as pollution, food shortage, or natural calamities, by arming the country with 'high tech'.[10] 'Crisis management' was first used to describe US foreign policy options during the Cuban Missile Crisis.[11] In the context of high technology industries, it refers to managing non-military events that could cripple a nation's domestic policy. Appropriate crisis management here centres on two important points: (a) the crisis/emergency is not enough to be handled in isolation, but needs to be integrated and co-ordinated with goals of other domestic policies. (b) Preparation for the crisis must be made well beforehand, and subject to constant update from new information.

Scarcity of natural resources has always been a concern of most ministries, especially MITI. In its *Vision for the 80s*, MITI places great importance on high technology, giving it the formidable task of 'conquering the restraints of a resource poor nation'.[12] There are several criteria for using high technology in stablizing supplies of basic resources, such as energy, raw materials, and food:

1. Updating plant and equipment as well as industrial processes to create more efficiency, promote storage and reduce the consumption of energy or mineral resources.
2. Developing new varieties of energy resources; producing new methods for regeneration of these resources; strengthening productivity of agriculture; and promoting animal husbandry and production of food.
3. Improving the exploration and collection methods for fishery and mineral resources.
4. Establishing sophisticated information networks for more effective resource management systems.

The importance of high technology in international relations

Since high technology has the potential to alleviate Japan's resource dependence on a few countries, the Council on Science and Technology (CST) has promoted a policy aimed at broader economic interdependence with other nations. The CST voices this line clearly concerning the contribution of scientific R&D to the international community:

> ... along with the elevation of Japan's economic and techno-logical power, her power of influence in the international society has grown immensely and also, the world's expectation of Japan has risen. This means that from now on, it has been demanded that Japan play an international role suitable to its position, not simply in economics, but in all fields including science and technology.[13]

This policy is a reversal of the still dominant foreign policy view, made famous by the early postwar leader Yoshida Shigeru, of confining foreign activities to a select group of nations, important to Japan in the supply of much needed resources. It represents a growing school of thought in Japan, known as the internationalists, which believes that Japan's maturity into an 'economic giant' concomitantly brings larger socioeconomic re-sponsibilities than before to foreign nations.

The Japanese have understood that the West will no longer tolerate a 'free rider' relationship. Articles reporting rumours of the US Congress being agitated over the number of US-educated Japanese PhDs returning home immediately upon graduation without contributing to American R&D, are symptomatic of the urgency of policy shifts.[14] Japanese newspapers are describing international interdependence as 'mutual existence, mutual pros-perity' (*kyōson, kyōei*) and 'a community bound by fate' (*unmei kyōdōtai*). (*Kyōdōtai* is an exceptionally apt metaphor since its historical meaning refers to small village communities tied by tight bonds of mutual dependency for survival.)

Recently, Japan has taken fresh action towards pursuing a practical approach to its often criticized position as an opportunist and 'free rider'. Until five years ago, Japanese companies were mainly concerned with buying marketing rights and licences for US-produced technology. Today, most of the policy initiatives have a different emphasis on co-operative research and joint development projects. Japan has pushed hard for collaboration across national boundaries. Summit Meetings have been eagerly used, for example, as platforms for voicing a willingness to participate in co-operative R&D projects that tackle challenging global problems. Cancer was a particular favourite of former Prime Minister Nakasone. In March 1983, he began planning 'a special request' to be made at the May Summit in Williamsburg, Virginia, for international co-operation in cancer R&D.[15] In 1985 in Tokyo, he proposed the 'Human Science Frontier Initiative' as a non-military alternative to the US Strategic Defense Initiative (SDI).[16]

High technology's role in promoting Japan's industrial production

Many Japanese companies, like their US rivals, have moved to high tech-based production because of a need to diversify into promising, new markets. Seiko watchmakers, for example, challenged by Hong Kong watch manufacturers – where 80 million watches are made each year at half the Japanese price – used their precision machinery to manufacture deoxyribonucleic acid (DNA) synthesizers.[17] Suntory, mainly a brewer of alcoholic beverages until 1980, established an Institute for Biomedical Research, and now produces interferon, monoclonal antibodies and tumour necrosis factor.

The Japanese government has carefully selected certain industrial sectors to provide supporting infrastructure for high tech-based industries. The CST recommendation of November 1984 states that 'the basis for the activation of the entire Japanese economy and for the development of the knowledge-intensification of (Japan's) industry is the activation of basic material industries and the advancement of process and assembly industries.'[18] The report lists five industrial areas, summarized below, necessary to boost Japan's industrial production capacity.[19]

Development and processing of new materials

Development of basic materials is 'one of the great motive forces' behind technological innovations for all industries.[20] This coupled with advances in working and measurement technology, such as high-degree, ultra-precise machinery capable of effective functioning at submicro- or nanometre levels, is important.

Process technology

New technology introduced into existing processes can: (1) enhance entire production systems; (2) be useful in integration and systematization of other processes and (3) create novel processes that connect raw materials and final product manufacture more directly.

Adding knowledge to production – management technology

Use of electronics and information technology aimed at maximizing efficiency and diversification of products.

Improving software productivity

In comparison with Japanese hardware capability, the reliability and productivity of software leaves much to be desired.

Regeneration and effective use of resources

In a world of finite resources, Japan, which is naturally resource deficient, must strengthen 'recirculation and practical use of resources'.[21]

The importance of high technology to regional development

The expansion of large cities, with the accompanying rapid urbanization, have made cruel demands on people living and working there. Overcrowding has rendered transportation by road inefficiently slow and by rail, hectic. Housing is poor by international standards – generally cramped, and highly overvalued.

In order to alleviate city congestion, MITI proposed in 1980 eighteen regional cities as technopolis. The Japanese word is derived from the Greek '*techne*' (arts) and '*polis*' (city) to portray the image of a city specializing in the industrial arts and applied science. The technopolis plan is primarily concerned with the development of medium-sized regional centres as potential locations of technical excellence, suitable settling places for the private sector. The industries (*san*) would rely on high technological innovations from the universities and national research laboratories (*gaku*), with researchers living in the city (*jū*). This three-part interrelationship consummates the government's much publicized *San.Gaku.Jū* plan for technopolis.

Not to be outdone by MITI, the Ministry of Posts and Telecommunications (MPT) devised in August 1983, their own version of a regional development plan called, 'teletopia'. As with 'technopolis', the Japanese word 'teletopia' is derived from two words of foreign origin: the English 'telecommunication' and 'utopia'. A document for internal use in MPT describes teletopia 'as a comprehensive policy for the smooth transition of regional societies into sophisticated informational societies based on new media.'[22] (New media is another popular buzz word, used often in the Japanese daily newspapers, and loosely defined as any revolutionary technology involved in the generation, collection, transmission, storage and processing of information. Examples include the teletext, a cross between a television and computer, and the videotex, a combination of a television, computer and telephone.)

Fourteen regional cities were named as possible sites for

teletopia. The completed system would link rural communities with quick and effective communication systems to other parts of the country. Like technopolis, teletopia is also an incentive for industry to move to a relatively isolated region. Both technopolis and teletopia serve another important purpose in economically igniting depressed rural areas. Farm, mountain and fishing villages have lagged far behind the swift pace of modernization set by *Tokaidō* (industrial sector) Japan.[23] Large migrations of people from country to metropolitan areas have deprived the rural community of essential human resources. Japanese regional policy encourages new technology to be used to restore some of the vitality, and to rectify glaring technological differences that continue to exist between cosmopolitan and rural areas.

Improvement of the quality of life and the general consensus for high tech

Improvement of social conditions through high technology is a policy area on which there is much emphasis. MITI's *1980 Vision* suggests that high tech could 'establish both national vitality and luxury'.[24] The eleventh Recommendation of the CST encourages the development of science and technology to 'build a more safe and secure society'.[25] In creating a high tech society, Japanese officials are privately apprehensive that mismanagement of technology could bring a repeat of the social and environmental disasters of the 1960s.[26] The lesson learned was to prevent a free reign of technology in the name of economic growth without regard to man's place in the environment.

Japanese high tech policy asserts that technology holds particular promise for a future Japan: in handling its ageing society, the need for a more relaxed living environment, and the demand for improvements in medicine and health care. Quality of life, usually associated in basic economics with having more buying power for consumption of goods, also means having more leisure time, and the freedom to enjoy favourite activities. High tech could add enormously to work efficiency allowing greater time for play. Home shopping, banking, health care systems, home-office links through automation, and home education opportunities are also benefits to be gained. Although these may be mere conveniences for most, high technology is imperative for the handicapped. At the Tsukuba Expo 1985, Nippon Telephone and Telegraph (NTT) displayed devices that seemed like obvious contradictions, such as telephones for those with impaired hearing. No matter how seemingly insignificant, high tech ideas can be integrated to give systems of much potential worth.

A 'late for the bus' mentality

Although expressed less often today, there was a broad consensus in the 1970s that Japan lagged behind the US in the development of new technology. This opinion is often supported by another popular view that US capabilities in basic research far exceed those of Japan. The US, for example, dominated global markets in high technology from 1970 to 1980 occupying nearly 40 per cent of the market share, while Japan accounted for a mere 5 per cent.[27] The US high tech R&D budget is still several orders higher than in Japan, and in some fields such as the soft sciences, the latter is said to be considerably behind in both R&D and the training of young researchers.[28] As late as 1985, for example, only one or two universities in Japan offered courses, not to mention an undergraduate concentration, in molecular or cell biology. In the US, however, thousands of students have graduated with first degrees in these subjects since the 1960s.

Running out of technology to import

Increasingly, there is an awareness that the 'Catching up Style of Modernization' that started with the Meiji Restoration has been supplanted by a Japanese superiority in certain technology. According to a survey in the 1983 STA White Paper on Science and Technology, more than 75 per cent of Japanese private enterprises indicated that their technological levels were equal or superior to their counterparts in the West.[29] This new confidence is not only with respect to technological parity with the US and Europe in capital-intensive industries, but also with respect to its ability to compete with the West in the new high technology-based industries. Dore suggests that the 1970s was a period marking a quantum-leap forward in Japanese technological self-confidence.[30] He continues that there was a 'growing realization' that Japan's 'competitive strength' lay in the spin-off of sound technology – quality, innovation and low costs of production design.[31]

The result of this popular agreement about the potential of high technology has been the public acceptance in Japan of the 'correctness' of government policy for its promotion. It is also important to note that in some ways the public is very concerned about high tech. Other STA surveys, cited in the STA White

Paper, have pointed to insecurity surrounding high tech and the relatively slow start of office and home automation. Yet, compared with many other countries, Japan seems quite receptive to new technology; a receptivity that is intricately tied with notions of high tech as a panacea for future Japan.

Is high tech a panacea?

Most laymen, lacking formal reading in the enormous literature on technical change in an economy, harbour many neoclassical assumptions about the role of science and technology in promoting economic development. Most of these assumptions, brilliantly articulated by Joseph Schumpeter in his influential *Business Cycles* published in 1939, regard science and technology as a mystical black box that can cure all societal evils.

Schumpeter explained the rate of technical change and its macro-economic consequences in terms of its relation with the rate of invention. Invention, Schumpeter argued, is exogenous to economic growth since it relies on the growth of science – also an exogenous factor to economic growth.[32] In this view, there must be an invention before there can be technical change. The practical application of an invention Schumpeter termed an innovation. Increases in innovative possibilities produce discontinuous spurts of economic growth. The growth is punctuated since innovations, in Schumpeter, are as unpredictable as mutations in natural selection depending 'on the random appearance of exceptionally gifted individuals'.[33] The neoclassical model for economic growth due to technical change can be summarized by the following linear projection:

SCIENCE-ORIENTED R&D → TECHNICAL PROCESS → ECONOMIC GROWTH

In Japan, as elsewhere, the laymen as well as most bureaucrats base their expectations of high tech 'miracles' for the economy on this linear model. Is the assumption that increased R&D leads to economic growth valid? Is Japan investing in the right fashion by re-allocating resources to develop basic research in pure science?

Since the 1930s, successive writers have rebutted and expanded the Schumpeterian thesis. Although the criticisms are varied, they agree that the basic flaw is in the model's simplicity; a straightforward linear progression of causality that ignores more complex interactions between science and technology, R&D,

invention and innovation, and economic growth. In order for a clearer picture of contemporary ideas about technical change, each step of the model will be considered.

Step 1. *Basic science R&D* ⟶ *Technical progress*

Mansfield in *The Economics of Technical Change* offers a useful definition of technology as 'society's pool of knowledge regarding the industrial arts'.[34] Schmookler further refined this definition by distinguishing between production and product technology; the former denoting knowledge used to improve the materials, machines and processes used in fabrication and the latter, knowledge used in creating the product itself.[35] Technology, then, is use-oriented knowledge; science is 'the sum total of systematic and formulated knowledge about the natural world'.[36] Science produces papers, technology does not. Since both science and technology represent two distinct bodies of knowledge, technology is not necessarily a derivative of science.

Since modern industrialized societies have had unique successes in applying systematized knowledge from R&D in their industrial practices, it seems that the driving force of technical change is an exogenous phenomenon. This has led to the view that the effect of basic science on technical change is as depicted in the linear model – a straight line causality:

> ... in the prevailing formulation of our time, it is common to look at causality as running exclusively from science to technology and it is common to think of technology as if it were reducible to application of prior scientific knowledge.[37]

It is well-documented, however, that technological knowledge for a long time had no reliance on science, and many powerful new technologies developed before there was systematized scientific knowledge of the phenomenon. Sadi Carnot's theories of thermodynamics, for example, were a result of working on steam engines; Joule's Law of the Conservation of Energy, was the product of experimenting with the power generators at his father's brewery. A. R. Hall writes about science and its effect on technical progress before 1760:

> We have not much reason to believe that, in the early stages, at any rate, learning or literacy had anything to do with it (technical progress), on the contrary, it seems likely that virtually all the techniques of civilization up to a couple of hundred years ago were the work of men as uneducated as they were anonymous.[38]

In this century, however, there is a general agreement that technical progress has gradually come to rely more on science. But, the literature still strongly suggests that, except perhaps for the new developments in the life sciences (discovery of DNA, recombinant DNA, and monoclonal antibodies), most technical breakthroughs did not depend a great deal on science.[39] The development of the computer, the first nuclear device, and heavier than air flight are all cases in point. Contrary to popular belief, new breakthroughs in technology are probably more important to the development of pure science than *vice versa*. The creation of the transistor, for example, in 1948 pointed to the importance of solid-state physics, and subsequent to 1950 there came a large-scale commitment of scientific resources in this area. It seems that the mutual dependence of science and technology is more deeply rooted than once appreciated and that science is certainly not an exogenous factor in technical progress. Rosenberg argues:

> I believe that the industrialization process inevitably transforms science into a more and more endogenous activity by increasing its dependence upon technology.[40]

In addition to the increased institutionalization of pure scientific research in private firms which forces a mutual dependency between science and technology, economic factors, such as the costs of R&D and expectations of financial rewards, also shape and direct scientific inquiry.

Step 2. Technical progress $\longrightarrow$ economic growth

There is much literature to confirm the role of technical progress in the development of economic growth. Moses Abramovitz in his essay, 'Resource and Output Trends in the US since 1870' and Robert Solow in his article, 'Technical Change and the Aggregate Production Function' support the claim that technology was important in the long-term growth of the US economy.[41] Technical progress is usually codified into either process or product innovations: the former concerned with improvements in the equipment or methods of product formation, and the latter with the modifications necessary to give the final product. The interaction between the two becomes very complex when equipment and machinery used in one process are end products of another. A more general definition of technical progress, accommodating both production and process innovations, is offered as any change in activity that produces from a given amount of resources either a greater volume of output or a qualitatively superior output, or both.

'Greater volumes of output' or 'superior output' have been called cost-reducing and performance-oriented processes respectively. Although the former is perhaps the main motivation behind most examples of technical progress, technology with high utility costs such as military and medical hardware have higher priorities on performance than expenses.

There are several difficulties, however, associated with the straight A to B causality that the linear model of technological progress to economic growth implies. The first is the near impossibility of separating other related factors of economic growth from technical progress; factors such as education, capital formation and resource allocation. In addition, there are the problems of not having unambiguous measures of output over long time periods, and the unsatisfactory methods used for assessing quality changes in products and their impact on an economy.[42]

Another important factor to technical progress is the effect of invention A to inventions B, C and D. Rosenberg sums it up:

> The social payoff of an innovation can rarely be identified in isolation. The growing productivity of industrial economies is the complex outcome of large numbers of interlocking, mutually reinforcing technologies, the individual components of which are of very limited economic consequence by themselves.[43]

This complementarity of inventions has been an important reason why some very spectacular technologies gave only gradually rising productivity curves. Historically, successful breakthroughs in technology awaited specific inputs from complementary activities before their full potential were realized. The compound steam engine, for example, had to await cheap, hard alloy steel resistant to high pressures and temperature. This high-quality steel was not employed, in turn, until the invention of new machine tooling to develop it. Again Rosenberg reiterates:

> Really major improvements in productivity therefore seldom flow from single technological innovations, however significant they may appear to be. But, the combined effects of improvements within a technological system may be immense.[44]

Hand in hand with complementarity of inventions is the importance of building a well-developed infrastructure for technical progress. This includes a means of facilitating education, capital formation, communication, the acquisition of natural resources and so on. Technological innovations will not realize their full potential without state-of-the-art communication tools, standardization of equipment, highly qualified scientists and engineers, and modern working facilities. A deficiency in any of these areas leads to longer lag times between the birth of a significant finding and the economic productivity it engenders. Indeed, the one factor governing increases in economic productivity associated with new technologies is that separate innovations never occur in a vacuum, but are highly interrelated, mutually reinforcing activities.

The Japanese bureaucracy's apparent faith in high technology, as portrayed by the mass media, policy statements, and government documents, is, in theory, based on ill-founded ideas about the interactions between science, technology, and economic growth. One simple, clear line of causality between basic science research and economic productivity just does not exist. In practice, however, the history of the relationship between science and technical change suggests that underlying the complex interactions of the linear model, other tendencies are set into motion which assure the overall trend towards economic growth.

First, a definite learning process, resulting from basic R&D engenders a series of small improvements in a system. Whether these accumulated modifications lead to technical progress or not (accumulated improvements in hot air balloons can never lead to passenger jets) is secondary to the knowledge created. Second, once small changes occur they diffuse very rapidly to affect other industrial sectors. The speed and frequency of this transfer of innovations to a new context increases the possibility that one improvement will affect technology in the same industrial sector or merge with complementary inputs from another industry to give economic growth. The very fact that the Japanese are fostering high expectations for high tech galvanizes entrepreneurial spirit and stimulates growth.

Japanese postwar history is full of examples of this competitive drive in entrepreneurship. In the postwar Japanese penicillin industry, for example, over 70 companies initiated production; today only three or four survive.[45] As complementary innovations of penicillin, many different designs of fungi cultivation tanks were produced. The improvement in tanks had an enormous effect on Japan's fermentation industry, and helped to build the Japanese reputation as world leaders of amino acid producers. This process is often compared to a giant electric generator wheel producing new ideas (electricity) as well as complementary innovations as by-products of the process (sparks). Many analysts,

59

such as Moritani Masanoru, believe that it is this spirit of competition that is the driving force of technological progress in Japan.[46] Finally, it must be remembered that reallocating resources in the Japanese economy and emphasizing the promotion of basic research is a political as well as an economic process. Different individuals, organizations and institutions have private reasons for fostering basic research that are not related to economic issues alone.

Uncertainty of high technology

A common problem in adopting new technology is the difficulty of assessing the long-term economic impact. Rosenberg asks the important question, 'Who in the 1770s, when James Watt was establishing the widespread commercial feasibility of the steam engine, could have anticipated that a major field of its application would be in new forms of transport, or that it would one day be used to generate electricity in order to illuminate houses?'[47]

The same underestimations followed many discoveries. The radio initially was thought to be only important for communication with remote areas where transmission by wire was not realistic. Commercial broadcasting opportunities, radio's use for entertainment, news and so on were almost totally unanticipated. Similarly, after oil was discovered in the US in 1859, it was sought not as a fuel, but as an illuminant since kerosene was critical for lighting. Only after the emergence of the internal combustion engine over a quarter of a century later did the oil industry appreciate the product's potential as a new power source.

Due to uncertainty over the use of new technologies, investment in high technology is often very risky. Since companies involved in promoting high technology cannot be expected to invest in products or processes of questionable returns, the government offers financial assistance – risk money. Risk capital is especially beneficial for firms conducting basic research with long gestation periods. Often, basic research, as we have seen, does not end in a new innovation, and to avoid unwelcome financial ruin of a large company, the government supplies the seed money and takes the losses.

Also, as noted, nations cannot engage effectively in basic research if supporting complementaries are not established. On-line computer data banks, collections of top information sources, standardization and regulation of products and processes for the safety of society are all important related activities. Private enterprises will not engage, however, in these endeavours since there is little profit incentive. Governments are thus responsible for providing the capital and expertise in establishing a productive environment for technological innovation.

Chapter Four

Strategies for developing biotechnology in Japan

In this chapter, a developmental view of the Japanese biotechnology policy process is taken. The nature of this process is more clearly understood if it is classically analysed as a series of sequential steps of policy demands, decisions, statements and finally policy outputs. In Japan, two periods separated by a gap of relatively little activity can be determined in the development of biotechnology policy. The first dates back to the early 1970s, when biotechnology was energetically promoted under the title 'life sciences'. Promotion of the life sciences was short-lived, and not considered vital to the nation's future economic well-being. The second period can be traced from the beginning of the 1980s and stretches to the present. During this time, biotechnology has enjoyed much publicity as a high priority area of tremendous economic potential receiving constant attention in the press.

The shared similarities of the different time frames in the promotion of Japanese biotechnology cast some light on the nature of Japanese policy-making processes. First, in both periods, policy demands are seen as claims by non-bureaucratic forces upon public officials for specific action. These forces are composed largely of the private sector or academics, but are not necessarily market related. In addition, although events abroad complicate perceptions of biotechnology policy demands in Japan, foreign activities had a direct influence on action that marked the beginnings of both the life science and biotechnology periods. Second, the final policy decisions are invariably limited to the Japanese bureaucracy. There is a virtual absence of policy demands for biotechnology promotion directed towards politicians. Third, in the two periods, policy statements made by government officials are reinforced by and reflected in both the mass media, or in advertisements and investment portfolios of individual business concerns. Fourth, both periods were galvanized into a 'boom' or peak of intensity by a rush of authoritative statements, extensive media coverages and general high public visibility. Finally, the policy outputs, even though a product of bureaucratic policy-making processes, usually reflected a large influence from those that demanded them; the industrialists and academics involved with biotechnology.

In all the ministries, specialized *Shingikai*, or advisory councils, have played key roles in structuring decisions about the development and implementation of biotechnology. Affiliated with the Science and Technology Agency (STA) and the Prime Minister's Office, there is the Council (*Kaigi*) of Science and Technology (CST); in the Ministry of International Trade and Industry (MITI), the Industrial Structure (*Shingikai*) Council; in the Ministry of Agriculture, Forestries and Fisheries (MAFF), the Technology (*Shingikai*), Council; in the Ministry of Education (MOE), the Science (*Shingikai*), Council; and in the Ministry of Health and Welfare (MHW), the Health and Welfare Science (*Kaigi*) Council.[1] In each ministry there is a complex, yet well-defined role played by each advisory council, the bureaucrats, and the academic/industrial consultant specialists not unlike that hypothesized by *Nemawashi*. Figure 1 shows these actors in a cyclical policy process of demands and policy action, with numbers describing the chronological series of events in policy formation.

Young, active researchers from universities and international businesses, with their bilingual skills are usually those that demand a particular policy initiative from a specific ministry (1). It is important here to distinguish the differences in status among Japanese researchers. The distinctions are usually drawn by age. Younger scientists are more active and aware of recent developments in their fields than their superiors. It is the senior researcher, however, who attends the executive meetings of the science societies or of *Keidanren*, Japan's Federation of Economic Organizations, and who communicates demands on behalf of all researchers in the academic or industrial world to the central government (2). The contacts between these senior scientists and the government are typically very informal, initially made through former students now working in the bureaucracy or by former bureaucratic leaders who have 'descended from heaven' to the company by *Amakudari*. Occasionally, a society or *Keidanren* submits reports to the ministries demanding action on an issue, but this is supplementary to informal letters, visits and telephone conversations. The demands generally carry the assumption that it is now the government's turn to act; usually along the lines of 'European and American researchers or industrialists are promoting

Figure 1 Simplified policy-making figure

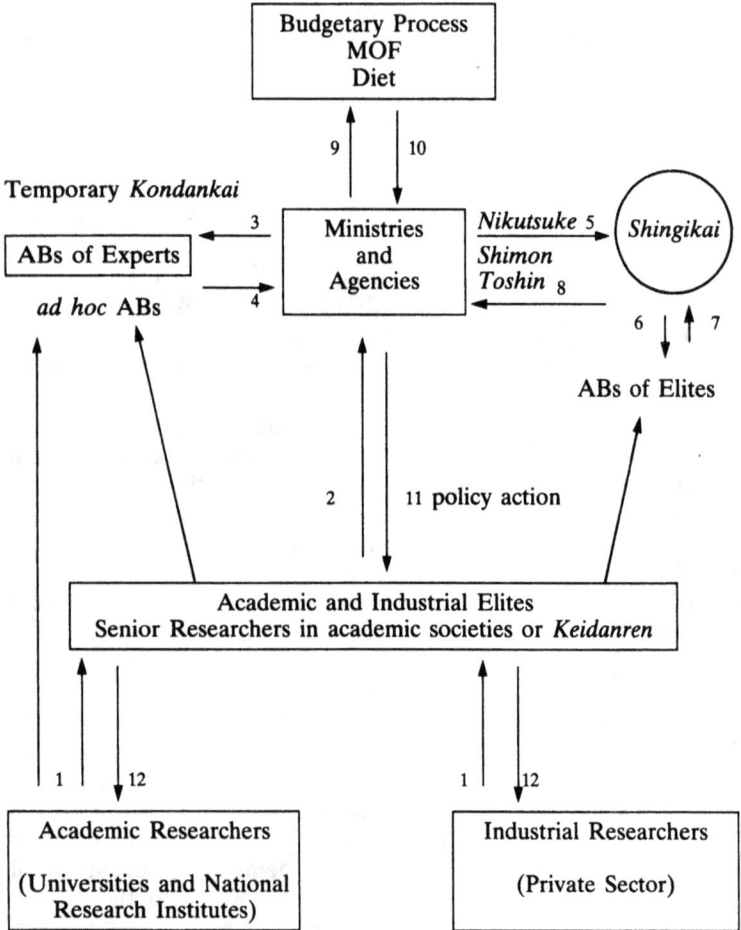

(1) Policy Demand from individual researchers. (2) Informal or policy demands (3), (4) Informal – by telephone, interviews, *Kondankai* meetings (5), (6) *Shimon* – Formal Inquiry submitted (7), (8) *Toshin* – Response with a formal report (9) Submission to MOF (10) Approved Policy (11), (12) Policy Action

N.B. often (3) and (4) are repeated later with most of the younger members of the *Kondankai* replaced

rDNA technology, we must do the same. It is your responsibility to do something.'

Action by the bureaucrats means massive data collection, forming an information clearinghouse to assess policy direction in a particular field. One of the quickest and most efficient ways to understand a new policy area is to be briefed by those who demanded it. The ABs formed to 'consult' the bureaucrats usually consist of a mixture of young, active researchers and their elite seniors (3), (4). The important point to note is the dual function played by these ABs. In Figure 1, for simplicity the ABs are labelled '*ad hoc* ABs of experts' and 'ABs of elites'. After the former fulfils its role as a bureaucratic consulting group, it is disbanded, and the bureaucrats use their advice to formulate the policy to near completion. The partially finished document is then forwarded to the advisory council in a process known as 'Inquiry' (*shimon*) (5). For the inquiry, ABs are reconvened, now with a majority of senior members present (6).

The senior specialists consulted are known as 'Boss Professors' (*Bosu no Sensei*), and their authorization of the policy (which was already drafted by bureaucrats) gives the document credibility as an advisory council report. The influential Tokyo University professor Watanabe Itarō, a personal friend of James Watson, the Nobel Prize winning discoverer of the structure of DNA,[2] is an example of a '*Bosu*' professor. In his career, he has served on myriad ABs advising on biotechnology policy (a personification of the new biotechnology academic *zoku*). Saitō Hyūga, also at Tokyo University and the Director of the university's Applied Microbiology Institute, is another well-respected researcher who has served on no less than six ministerial committees related to biotechnology.

Committees of well-respected 'Boss Professors' established to endorse a policy are often referred to as *tatemae* committees, since the bureaucracy falsely acknowledges the committees' 'Responses' (*Tōshin*) as objective, independent opinions. These committees, however, act as *tatakidai* or sounding boards for their affiliated ministry or agency. The activity of endorsing a policy draft to expedite its passage through the budgetary process is known as *Nikutsuke* – putting meat on a proposal. The policy-making process in Figure 1 is depicted as circular because once a report is approved by the Ministry of Finance (MOF), the immediate benefactors are the university professors and in-dustrialists who initially advocated it.

It is often difficult, for example with an Advisory Council such as the CST, to determine exactly what input has come from the bureaucrats through (5) of Figure 1 and what influence has derived from the specialty subcommittees (7, 4, 8). With the roles of the civil service, advisory councils and academic specialists in mind, let us examine the historical development of biotechnology policy in Japan.

Development of the life sciences in the 1970s

The first stage of what was to become 'biotechnology' began roughly in the early 1970s with the growth of the life sciences. Before this, most of the R&D associated with the biological sciences was restricted to decentralized activities in universities and separate national research institutes. Like other examples of reallocating resources in the postwar Japanese economy, the central government, more specifically the Prime Minister's Office, was responsible for focusing activities, and organizing R&D. During this development period, four main bodies took centre stage: the MOE, the CST, the STA and *Keidanren*. The Japan Science Council (JSC), a permanent AB associated with the MOE was peripherally involved particularly in the early stages.

In the mid 1960s, the JSC, through its National Committee for the Biological Sciences demanded that the MOE formulate a more clearly defined national strategy towards the biological sciences. It was about this time, however, that the JSC suffered a gradual undermining of its credibility as a *bona fide* permanent AB to the MOE. The organization was crippled by a growing radical left-wing, and increasingly became a platform for criticizing government science policy. Needless to say, in the 1960s the MOE did little to follow its advice, and in an attempt to constrain its power, created the Science Council (*Shingikai*) – today's nerve centre of the MOE's efforts to promote biotechnology.

It was not, therefore, until the early 1970s with the efforts of the CST and STA that Japan began formulating a comprehensive life science policy. These two bodies have consistently worked together throughout the 1970s and 1980s in the life sciences/ biotechnology policies. The general *modus operandi* is for the CST to announce a policy initiative (usually in its 'response' to inquiries initiated by the Prime Minister), and for the STA to follow up the policy proposal with appropriate action. This system of the STA heeding the advice of the CST before implementing action is often dubbed *gogisei* or a council system. It is probably due to both the close structural ties between the STA and the CST and their joint location in the Prime Minister's Office that a functional symbiosis has evolved.

Strategies for developing biotechnology in Japan

The more important of these two bodies was the CST. It was this organization that emphasized the significant role of the life sciences as part of a broader strategy for promoting knowledge-intensive industries in a report (*Tōshin*) on science and technology in April 1971. The *Tōshin* of the CST was a 'response' to a specific 'inquiry' (*shimon*) demanded by the Prime Minister on 25 August 1970.[3] The rough response was drafted by the various subcommittees of the CST on 15 April 1971, and presented to the Prime Minister less than a week later as a final report. Although this process hints at the normalcy of an AB under an organizational framework of *Nemawashi*, because the CST has become the most powerful and institutionalized of all the permanent ABs in science, its responses or recommendations are closely monitored by STA bureaucrats.

The CST was set up as 'an advisory body to the Prime Minister, based on the CST Statutory Law of February 1959, to be instrumental in the comprehensive promotion of the nation's science and technology policy'.[4] It consists of ten members, excluding the chairmanship held by the Prime Minister, usually divided equally between politicians and non-politicians alike. (Invariably there are seats for the Finance Minister, the Education Minister and for the Directors of both the Economic Planning Agency and the STA.) In the daily newspapers the CST is referred to as 'the highest strategic body of science'.[5] The recommendations and opinions (*Iken*) it submits to the Prime Minister are effectively *de facto* national policy guidelines towards science. These CST reports, of which there have been eleven at the time of writing, are very influential, and differ enormously in importance from the myriad reports of less-permanent subcommittees and *kondankai* committees in the central government.

After the initial report on life sciences in April 1971, the CST began a more focused approach to the promotion of these sciences. In 1972, as a Minister of State for Science and Technology, Nakasone Yasuhiro, remarked, '... from the standpoint of one who administers science and technology, it (a true equilibrium of science) may be defined as restoring the balance between the biological sciences and the physical sciences and engineering, for the former is lagging behind the latter.'[6] In May 1972, a Life Science *Kondankai* was formed by the CST, and after the submission of two reports in September and December, it was 'developmentally restructured' – promoted – to a permanent AB on 9 July 1973. This committee, the Life Science Panel, with 20 members of whom six are in the CST, is a private

AB to the STA Director General and remains the backbone of the CST's life science policy efforts today.[7]

The general excitement created by the 'talk' and policy statements about the life sciences in the mid 1970s bolstered its importance. Although the official duties of the CST's Life Science Panel were to be 'concerned with laying the policy foundation to establish long-term, comprehensive research goals in the life sciences and to promote necessary research, particularly important research, to achieve these goals,' its immediate role was not so high sounding.[8] It served primarily to prompt the bureaucratic organization closest to it, the STA, to recognize the importance of the biological sciences to the knowledge-intensification of Japanese industry.

Before the STA initiated any action towards the life sciences, *Keidanren* responded in 1973 to the policy statements of the CST. *Keidanren* respresents big business in Japan, and as of July 1986 had over 875 corporate members.[9] Founded forty years earlier in August 1946, it boasts an elaborate system of standing and special committees, ABs, as well as less formal groups; an advisory system that has enabled *Keidanren* to respond to government requests and formulate its own statements and recommendations voicing its preference on specific issues. In maintaining this tradition, *Keidanren* launched the '*Kondankai* (Committee) About the Life Sciences' in November 1973, and after its first major report entitled, 'An Opinion About the Promotion of the Life Sciences', promoted the *Kondankai* to a standing committee in December 1974. In 1973, most companies gave little more than cursory glances to *Keidanren's* early initiatives in the life sciences. Interestingly, *Keidanren's* response to official and unofficial cues from the government at that time was incited by a government-initiated bioboom. In 1980, however, the situation was reversed since it was *Keidanren's* active participation that helped to prompt a second biotechnology mania.

The STA was the only organization in the central government that replied to the CST's efforts to develop the life sciences in the mid-1970s. The STA is in charge of the administrative affairs of all five advisory organs to the Prime Minister's Office – the CST, the Space Activities Commission, the Council for Ocean Development, the Atomic Energy Commission and the Nuclear Safety Commission.[10] In 1974, the agency established an Office for the Promotion of the Life Sciences in its main R&D laboratory, known by its Japanese acronym, RIKEN (*Rikagaku Kenkyūjo* – The Institute of Physical and Chemical Research), in Wako City, near Tokyo. The STA's sensitivity to the CST's recommendations was clearly evident in 1974, and has become even more pronounced today.

The STA, as outlined in the Statutory Law in 1956, has a special responsibility for generally coordinating administrative works performed by all central government offices in science and technology. In addition, the agency functions as a normal ministry involved in planning, formulating and implementing policy in its specific jurisdiction. The life science case highlights how the STA has begun to disregard its role as a general coordinator of Japan science policy, the assumption of this forsaken role by the CST and the adoption by the STA of a more active part in normal ministerial activities. Already it has been observed that the CST took initial responsibility in the 1970s in planning and formulating basic policy in the life sciences; the STA was content to wait three years and respond with an implementation strategy to promote R&D at RIKEN. In 1977, the STA continued its activity in implementation by launching six new life science related projects at RIKEN. Ironically, while the STA was initiating this R&D, the CST persisted in its efforts to coordinate policy in the life sciences by publishing its sixth response, 'About the Foundation of a Comprehensive Science and Technology Policy on Long-term Prospects' in May of that year.[11]

Also in the late 1970s, both the government and the private sector in Japan, seemingly oblivious of developments in the US, did little to promote the life sciences. The exceptions were the activities of the STA, the MOE, and the CST who encouraged regulatory policies towards rDNA. This lack of interest quickly changed when biotechnology suddenly emerged in 1980. Indeed, even the CST's Life Sciences Panel, which published an interim report on the life sciences on 11 December 1974, did not think it necessary to submit its final report until six years later, 14 August 1980.

This lag in activity was caused by several factors. First, the life sciences, especially rDNA technology, were still very much in an experimental stage, carried out mostly in university laboratories, and thus the jurisdiction of one ministry, the MOE. Large-scale industrial prospects were very distant. Second, industrialization of rDNA technology was an horrific thought to most during this period. There were exaggerated fears of man having the ability to build human clones, or to create science fiction-type superbugs capable of destroying himself and the environment. Memories of *Minamata Byō* and *Itai Itai* (literally Ouch ouch) disease, sicknesses caused by mercury poisoning in the waters around

Minamata Village, Kumamoto Prefecture, were still too fresh to allow for irresponsible policies towards the environment. The consensus in Japan was to wait until there was enough information and confidence for industrialization to begin, which generally meant in more practical terms, to wait and see what the Americans would do. Third, so little was known about the new technologies of the life sciences that both the private and public sectors deemed them too economically risky to reallocate scarce resources especially during the country's recovery from the 1973/ 74 oil shocks.

MITI as flag waver

In Japan, newspapers often label 1980 as biotechnology's *gannen* (original year). Despite the lack of involvement of MITI in the 1970s and the leadership role taken by the CST and the STA up until the end of 1981, it was really MITI who thereafter gave Japan a *de facto* co-ordinated and systematic approach to 'target' biotechnology-based industry as a high priority sector. This role of MITI, even in science policy, to incite a transformation of precious resources to another industrial sector of the economy has been referred to as the '*Hataburiyaku*' – the role of a flagman, a starter or initiator of policy.[12]

The STA and the CST were reluctant to relinquish the well-earned authority they had mustered in the 1970s as co-ordinators of life science policy in Japan. Indeed, even the drafters of the January 1984 OTA Report commented on the STA's past significance and present strength, 'Until MITI's entry into major biotechnology programming in 1980, the STA's R&D programs in fields related to biotechnology were the largest and the best funded in Japan. Even today, the STA's programs are comparable in scale to those of MITI.[13] In fact, it seems that throughout 1980 and 1981, both the CST and MITI were involved in a *Nawabari Arasoi* struggle for ultimate influence in controlling the direction of biotechnology policy in Japan; a struggle that MITI won.

Other ministries within the bureaucracy, such as the MAFF for example, had been engaged for some time in decentralized, loosely structured activities in biotechnology such as the collection of plants and seeds in numerous regional research institutes. This practice, however, was not only low in priority but poorly funded. It was not until MITI's efforts of 'targeting' biotechnology as a future mainstay of the Japanese economy that serious commitments to plant collection and gene bank development were made by the MAFF.

Strategies for developing biotechnology in Japan

MITI's policies towards biotechnology, and the bioboom that followed in the autumn and spring of 1981–82, can be interpreted as the results of parallel courses of development. On the one hand, biotechnology policy strategy can be analysed as a chronological promotion of policy demands precipitating policy decisions, statements and finally action. Seen in this way, the 'targeting' of biotechnology in Japan coincides with a series of domestic and international events in the autumn of 1980, the most significant of which were the activities of new biotechnology firms (NBFs) in the US. This Japanese reaction to US activity was an important impetus for the development of biotechnology policy, but does not give the full picture. On the other hand, if biotechnology policy is viewed not as a result of a 'policy-demand-incites-action' styled decision-making process, but as a response to MITI's *Vision of the 1980s* to combat structural problems in the Japanese economy, then the 'targeting' of biotechnology can be dated much earlier. Both of these parallel courses are examined in turn.

Policy-demand-incites-action theory

In October 1980, Genentech, the largest of the American biotechnology concerns, set a record for the fastest price/share increase ever registered on Wall Street, with shares more than doubling from $35 a share to $89, in less than 20 minutes.[14] On 11 October 1980 shortly before the record-breaking equity offer, Robert Swanson, the president of Genentech, visited Tokyo to address the 1980 Microbial Industrial Technology Symposium at the *Nihon Seinenkan* where he described Genentech as a company fast becoming 'a second Sony' through the marketing of genetically engineered products.[15] Then, several weeks later in early December, the Cohen–Boyer patent for rDNA technology was issued in the US sending warning signals to the Japanese who feared that the patent would affect any product of genetic engineering, and prevent the benefits of US biotechnology research from flowing to Japan at nominal costs.[16] This Cohen–Boyer decision had been prefaced by a ruling, after eight years of litigation, by the US Supreme Court on 16 June 1980 allowing a General Electric Company's scientist, Dr Ananda Chakrabarty, to retain a patent for a genetically altered (not with rDNA technology) strain of oil-degrading *Pseudomonas* bacteria. Although the Chakrabarty decision acknowledged the right of private ownership of microorganisms, Stanford University's Cohen–Boyer patent marked a change in US policy for permitting universities to procure patents for federally funded research inventions.[17] The rapidity of the Supreme Court's ruling on the Stanford patent caught the Japanese private sector unaware.[18]

On 12 December two weeks after the patent was issued, an emergency meeting of *Keidanren's* Committee on Life Science was called to respond to the new developments in the US. Morishita Noboru, a journalist for *Nikkan Industrial Newspaper*, described the tense atmosphere as 'Japan on the brink of an international patent war', and the surprise move provoked accusations by business leaders of the US plotting a 'Reverse Pearl Harbour'.[19] After the meeting, Kyōwa Hakkō's Katō Bensaburō, the first chairman of the *Keidanren* Committee, remarked on the change in mood at the emergency meeting, 'It's been different up until now, today was a day in which the committee virtually launched a new business.'[20] The *Keidanren* (Committee) reportedly hastened into 'combat readiness' and warned both MITI and the STA that the American NBFs were moving biotechnology from a decade of experimentation into an industrial level of development; all of a sudden commercial prospects such as genetically engineered human insulin and interferon were real.

Bioboom

There was no question that by the beginning of 1981 the US biotechnology craze had been 'imported' into Japan. The excitement of genetic engineering fever hit the Japanese stock market almost overnight. The press organized a winter and spring offensive on biotechnology.[21] Japanese chemical and pharmaceutical firms rushed into the new industry. As early as 14 October 1980, Green Cross unofficially announced that it would begin a technology agreement with an American company in the spring of 1981. Takara Distillers, formerly an old distillery in Kyoto, had young researchers combing the Sahara in Algeria and the terrain of Papua New Guinea in search of industrially viable bacteria. *Keidanren*, in May 1981 'promoted' their temporary Life Sciences *Kondankai* to an *Iinkai* – a committee with more permanent status as an AB (see Chapter 1). Rapidly biotechnology was gaining a concensus among the *Nebukai* coalition members as an important area of high growth.

An indicator of the degree to which positive imaging of biotechnology was used by the Japanese press was evident in 1984 when Kanebo Pharmaceutical's 'Bioseries' of lipstick was marketed. The lipstick's pigment, shiokinin – normally obtained from the roots of the plant, gromwell – was produced by culturing gromwell roots in 750 litre bioreactors. Sales of the lipstick were phenomenal in Japan, over three-quarters of a million a month. The lure of using a product derived from biotechnology was the main attraction.[22]

MITI responded to these cases of biofever by proposing in its budget for FY 1982 the establishment of a Bioindustry Office along with an AB to support it, the Bioindustry Advisory Committee (BAC). MITI, however, was determined to put its own indelible stamp on the direction the industry turned. It viewed biotechnology as a convenient catalyst for diversifying and reviving the Japanese chemical industry which had been suffering from over capacity since 1978. Even the locus of the Bioindustry Office in the Basic Industries Division of MITI, that part responsible for steel, non-ferrous metals and chemicals, reflected MITI's initial efforts to use biotechnology policy as an opportunity to adjust struturally the failing chemical industry.

MITI also considered biotechnology, at least in the initial stages, as part of a larger strategy for developing future energy sources, and for reducing Japan's dependence on imported oil. It established a Biomass Policy Office in 1980 and began several projects including the R&D consortium, the Research Association for Petroleum Alternatives Development (RAPAD). On 1 June 1982 the Biomass Policy Office underwent 'developmental re-structuring' (*hattenteki kaisō*) forming the Bioindustry and Energy Policy Offices. The functional similarities of these two offices justified their establishment in the same room separated by dividers. There has been a recent trend, however, with rapidly decreasing oil prices and Japan's success in diversifying energy sources, for MITI to de-emphasize its concern with the energy-saving potential of biotechnology. The Energy Policy Office, for example, has been greatly reduced in size occupying only a corner of the Bioindustry Office.

BAC, the AB set up to advise the Bioindustry Office, was formed a mere month after the Office itself. It was a private advisory body to the director of the Basic Industries Bureau of MITI with the specific task of producing a future vision for Japanese bioindustry – quickly dubbed by the press, a 'Biovision'.[23] Dr Saito Hyuga, chairman of BAC as well as the head of Tokyo University's Applied Microbial Research Institute, described the role of the Bioindustry Office and his committee as:[24]

1. promoting R&D
2. producing a future vision for bioindustry
3. drawing up plans for aiding bioindustry by promoting international co-operation

Biotechnology in Japan

4. safeguarding and nurturing bioindustrial results usually with patents

MITI's establishment of a Bioindustry Office was also part of a larger co-ordinated effort by MITI to obtain special legislation for the promotion of Japanese biotechnology as there was in the electronics industry – the 1957 'Extraordinary Measures for the Promotion of the Electronics Industry'. This attempt by MITI failed – no special law has been passed.

The next generation basic technology project

Although MITI was a flagwaver in initiating a centrally co-ordinated bioindustrial development strategy, Japan's biotechnology efforts had already been in full bloom by the time the Bioindustry Office was established in June 1982. Even though MITI's actions in the summer of that year set important precedents in policy initiatives for other ministries who rallied in the next two to three years to 'cash in' on biotechnology, there was another MITI influence that had acted earlier – a *de facto* policy towards biotechnology that gave much-needed direction to Japan's private sector. This *de facto* policy also sent international shock waves throughout the world inciting swift reactions from foreign governments determined not to allow the Japanese to overshadow them again economically. Ironically, this policy was not a result of demands from the private sector for deliberative policy action towards biotechnology promotion, but rather a mere R&D implementation response to MITI's *Vision of the 1980's* advocation for a structural adjustment in the economy.

This MITI policy was the establishment of the next generation basic technology project (NGBT), the result of a parallel course of development almost totally separate from the 'policy-demand-incites-action' styled decision-making process described previously. In order to trace the development of NGBT, it is necessary to examine the origins of MITI's *Vision of the 1980's*. This analysis is important since it highlights how a powerful Japanese bureaucracy drafts the outlines of an influential policy document.

The 1980 MITI Vision

The *Vision of the 1980s*, published on 17 March 1980, was MITI's first attempt at a general, co-ordinated policy towards the high technology industries of the 1980s. Like the *Vision of the 1970s* before it, the *Vision* outlined a prudent path for Japanese technological development for the next decade. Unlike the 1970 version which proved outdated months after publication – initially due to the Nixon shocks and later as a result of the oil shocks – the *1980 Vision* was very well received. It is often quoted in the introductions to reports by ABs formulating biotechnology policy in MITI as the guiding policy document of the 1980s. MITI's Agency of Industry Science and Technology (AIST) described the main thesis of the *Vision* as 'to grasp changing political, social and economic conditions in the world as well as the diversified needs of our nation in the 1980s and to set forth a basic framework for the formulation and implementation of our industrial policy in this decade'.[25]

In addition to its influence in the bureaucracy, the *Vision of the 1980s* has had a large impact on the Japanese private sector. A former MITI Planning Office official, remarked, 'What the private sector probably wants from MITI is to be shown in the international trend of things which direction in general is good to move in. What MITI does is wave its flag and say, "This way is probably the best."' The era of possessing coercion over the industrial world was over ten years ago.'[26] Akazawa Shoichi, a Fujitsu vice president and a member of the biotechnology subcommittee set up to authorize the content of the *Vision of the 1980s*, was uniquely qualified to assess the function of the '*Vision*' for the private sector. He suggested that 'Since the 1980s has more uncertainty than the 1970s, it is natural for a Vision not to say anything very concrete or detailed, but be overall a "positive" thing.'[27]

As outlined in Chapter Three, the main theme of the *Vision* calls for 'the establishment of economic security' in order to create 'a nation built on technology ... to conquer the restraints of a resource poor nation'.[28] This can only be achieved, it is argued, if Japan actively promotes new innovative technology, primarily microelectronics and information technology, biotechnology, and new materials. Drafting of the report was handled by a permanent, institutionalized *Shingikai* in MITI, the Industrial Structure Council. Established in 1964, this deliberative council consists of industrialists, financiers and bureaucrats who provide a joint public–private forum for producing long-range industrial policy documents. Although this permanent AB is serviced by a variety of specialized subcommittees chaired by capable academics, most of the initial reports, writings, recommendations, and notes of the Industrial Structure Council are formulated by MITI bureaucrats. In the case of the influential *Vision for the 1980s*, especially, the Industrial Structure Council and its various subcommittees were examples of *tatemae* (fronts) used to make popular and authorize policies of MITI.

Formulation of the Vision

Since the *Vision of the 1970s* is generally regarded as having fallen short, the first question asked by MITI was whether or not a *Vision* for the eighties was necessary. All the formative work was done by MITI officials. From February to May 1978, a feasibility study was carried out by an in-house committee, the Ordinance Investigation Committee, that concluded, 'A strategy map different to that of the 1970s was essential.'[29] This feasibility study assessed, in effect, what private enterprises and academic specialists at home and abroad were thinking about high technology-based industries. Much of the investigation was done informally by telephone or by young bureaucrats arranging personal interviews with influential industrialists. As Ikegaya Soichi suggested, 'Few businessmen would dare deny a bureaucrat an audience.'[30]

On 25 July 1978, twenty five young bureaucrats, mostly section heads (*kachō*) and assistant section heads, established the Preparation of the 1980 Vision Industrial Committee. In two months they had formulated a working framework for the Vision entitled *Issues and Topics*. After this report was distributed internally in MITI, a formal intra-ministerial investigatory body (*kentō kikan*) was set up on 26 October 1978. The body, 'The 1980 Vision MITI Policy Research Committee' was the main organization responsible for creating the Vision. It had seven subcommittees (*bunkakai*), each headed by a section head and served by approximately twenty bureaucrats of the assistant *kachō* or related class giving a total number of 144 MITI officials involved. An additional 100 non-governmental 'experts', largely industrialists and academics with a long-time involvement in these issues, participated in nine special *ad hoc* groups established for free debate of 'hot' topics. Since the Ordinance Investigation Committee also continued to meet, at any one time there were nearly twenty committees collectively involved in compiling the *Vision for the 1980s*. The final draft was produced for internal use in MITI on 24 August 1979. As Dore explains after interviewing MITI officials and subcommittee members concerned, 'What is certain is that the August 1979 draft was treated within MITI , where it circulated freely, as to all intents and purposes the final version. As such it soon acquired scriptural status.'[31]

On 24 August 1979 MITI also formed a *tatemae* biotechnology subcommittee chaired by a respected Tokyo University economist. The committee's role was explicitly to help draft and evaluate the *Vision of the 1980s*, but implicitly to sanction the draft of the Vision just completed by MITI bureaucrats. The 1980s Policy Subcommittee, as it was known, was under the General Purpose

Committee – the main committee of the Industrial Structure Council – and consisted of eleven businessmen, eight professors, three economists, four journalists, one novelist, two trade unionists, two representatives from women's organizations and six representatives from organisations such as the national chamber of commerce.[30]

Foreign observers and journalists unfamiliar with Japan often mistake committees such as the 1980s Policy Subcommittee as the backbone of the Industrial Structure Council in their haste to laud the Japanese government's use of objective, 'independent' ABs in formulating an important policy document. More cautious analysts, even if they reject 'the objective AB line', point to the importance of a subcommittee of wide member diversity as support of consensus building in a *Nemawashi* type concept. As far as the MITI bureaucrats were concerned, the *Nemawashi* perspective had already been applied in achieving an agreement among the approximately one hundred 'experts' on the nine special *ad hoc* committees to whom the officials turned for reference. The 1980 subcommittee, however, was largely a *tatakidai* (sounding board), or legitimizing body for the *Vision of the 1980s*; as the Director of the AIST's research department was reported to have responded to a policy question while handing over internal 'guideline' drafts of the *1980 Vision* to the subcommittee in October 1979, 'Well, the way the matter is handled in the *Vision* . . .'.[33]

Around the time MITI began work on the *1980 Vision*, the AIST had already started formulating a working outline of the NGBT project. In fact, as early as 1977, the AIST formed the Committee for Long Range Plans for the Development of Industrial Technology – an *ad hoc*, private non-statutory AB to the Director of the AIST. The chairman was a Tokyo University professor, and the other members consisted of ten university professors, two scientists at research institutes, a newspaper reporter and eight people from private industry. For the AIST, the committee seemed to serve three functions. Initially, the committee genuinely appeared to be an information channel for opinions and plans to develop a project suited to the 1980s. Dore remarks that the MITI officials he interviewed 'readily acknowledged' what they had learned especially from the young sectoral specialists.[34] Second, he also adds that the committee was a convenient policy tool of the AIST to procure a budget through which it could commission surveys to assess industry's opinions.[35]

The third function was one of *tatemae*. After the AIST officials had exhausted the informational usefulness of the 1977 committee, it was abandoned, but not the ideas it engendered. An internal MITI committee of section heads reformulated the working framework of the project. A second committee of bureau chiefs then rubber-stamped the real work already done and passed the project to the 1977 Long Range Development Committee for lay member approval.

It is important to realize that the bare bones of both the *Vision* and the NGBT project were drawn up and generally accepted within MITI at the same time – about the autumn of 1979. These policies were compiled by sharing the enormous volumes of surveys, reports, evaluations, statistics and notes on hearings or interviews generated between them. Within MITI, the Promotion Bureau was responsible for facilitating co-operation and sharing of this wealth of information between officials of the AIST, and bureaux such as those of Basic Industries, Consumer Industries and Machine Information Industries.[36] Given the sense of obligation each bureau felt to its particular industrial sector, and to the leading firms involved, efforts were made to provide each bureau and its particular sector with a fair share of the benefits of the *1980 Vision* and NGBT project. With this *Nemawashi* tenet in mind, the decision was made to designate microelectronics (Machine and Information Industries Bureau), biotechnology (Basic Industries Bureau), and new materials (Consumer Industries Bureau) as high priority areas in the *1980 Vision* and to co-ordinate their implementation through long lead-time, risky basic research projects in the NGBT programme.

The question still remains, however, 'What was there in the huge volume of material which promoted the internal MITI decision to sponsor biotechnology?' Dore is probably closest in suggesting that the information consisted largely of foreign literature, since the Japanese felt as though they lagged behind Western nations in biotechnology.[37] Again the government's use of the word '*ōbei*' as described in Chapter One, compared Japanese development as behind the US and Europe. In accordance with the *Nebukai* perspective, a distinction must be made here between the Japanese interpretation of the situation abroad and the actual events that transpired. The Japanese perception of the US picture, for example, was somewhat accurate – NBFs were rapidly forming in response to opportunities in biotechnology.

In Europe, however, the Japanese interpretation of events, as portrayed in the press and government policy statements, was grossly misleading. West Germany, for example, with a traditionally strong fermentation industry along with solid pharmaceutical and chemical sectors – a natural for biotechnology – was seen to be particularly formidable. Although as early as 1968, DECHEMA (*Deutsches Gesellschaft für Chemisches Apparatewesen*, a chemical plant association), provided the initial stimulus for biotechnology development, it was not until 1980 that West Germany shifted its R&D priorities to rDNA technology. Like Japan, the Germans are weak in rDNA research. In the 1970's, Germany remained locked into a preoccupation with second generation bioprocessing epitomized by the giant chemical company Hoechst signing in May 1981 a $50 million deal with the Massachusetts General Hospital to access American know-how in rDNA.

Similarly with Britain, the Japanese government and press commended the UK as the first country after the US to develop rDNA guidelines in August 1976. The implication was that Britain had a well-developed, centralized strategy in rDNA technology. This was very far from the truth; all the guidelines indicated was that the UK paid careful attention to researchers conducting rDNA experiments, not uncommon in a country that prides itself on the high quality of its basic research in the biotechnology-related sciences. In March 1980, Britain released the *Biotechnology: Report of a Joint Working Party*, known as the *Spinks Report* on Biotechnology, convincing the Japanese again that the Europeans were 'targeting' the new technology.[38] Ironically, the Spinks Report generated more excitement in Japan than in Britain; the Department of Industry casually acknowledged the report a year later with a twelve-page 'token' White Paper.[39]

Lobbying

After developing a plan to bring Japan to par with the rest of the industrialized world, MITI faced a final hurdle; lobbying the NGBT programme through the MOF. The NGBT project received a very favourable response from the MOF for several reasons. First, the programme's concern for the long-term development of basic technology in Japan was 'the quintessential embodiment' of the 1980s *Vision*.[40] Second, in the autumn of 1980, a biotechnology *Kondankai* of five companies was formed as a lobbying group to promote the programme's chance of success. The companies involved were Mitsubishi Chemical, Sumitomo Chemical, Asahi Chemical, Kyōwa Hakkō and Mitsui Toatsu. These firms later formed the core of a research consortium that implemented the NGBT project.[41] Third, in August 1980 the AIST Committee for the Long Range Plans for the Development of Industrial Technology published its interim report reiterating the importance of R&D in basic technologies in

Japan. (The fact that this committee, which was one of the first committee's responsible for the NGBT project, published an interim report after the project was to all intents and purposes completed, is further evidence of its perfunctory role.) Fourth, perhaps the biggest boost towards 'selling' the NGBT programme to the MOF was the Genentech equity flotation boom. Ironically, this event was not orchestrated by MITI, but occurred because of changing international economics. MITI here is similar to the scientist, who on falling upon a miraculous cure by serendipity, is praised by his colleagues not for his good fortune, but for his 'scientific preparedness' in exploiting a rare opportunity. This is consistent with MITI's philosophy expressed by former minister Konaga Keiichi, 'MITI is an opportunist; it doesn't back losers.'[42]

As a result of the biotechnology fever hitting Japan, the biotechnology aspect of the NGBT project received a total financial allotment of ¥26 billion to be spread over ten years (the first year's budget alone was ¥675 million). By 1 August 1981, fourteen companies were organized into a R&D consortium. Three companies formed a team to investigate rDNA technology, six for researching bioreactors, and five for examining mass cell culture. The structure of the NGBT consortium also mirrored MITI's policy of revitalizing the chemical industry. Of the fourteen companies 'contracted' to participate in the biotechnology aspect of the NGBT project (all references hereinafter to NGBT are considered to be towards the biotechnology part of the programme), eleven were related to the chemical industry.[43] Moreover, the public announcement of the research association on 11 August 1981 was made at a meeting of the Japan Chemical Industrial Association presided over by the head of the NGBT project Suzuki Eiji, who was the president of one of Japan's largest chemical concerns Mitsubishi Chemical.[44]

The NGBT project enjoyed high public visibility at home and abroad through glamorous news coverage. A *Nebukai* analysis suggests that foreign governments and companies alike feared the NGBT project because of its status as a research consortium with the potential of harnessing the co-operative power of 'the Japanese' like other consortia before it, such as the celebrated VLSI (Very Large-Scale Integrated Circuit) project (see Chapter 5). The NGBT's popularity at home was not so much due to its status as a research consortium, but to its function as the government's first co-ordinated initiative towards biotechnology development; an initiative that effectively created a national consensus on how to tackle biotechnology. Although the NGBT was only an indirect result of a demand for action precipitated by the

Genentech stock fever, and a fear of losing access to foreign patents, this was forgotten even by key industrialists involved. The project leader, Suzuki, exclaimed after the NGBT programme was announced, 'At this rate, the important patents will be rightly controlled by the US/Europe (*Ōbei*) without Japan moving an arm or a leg. At least by bringing together the government and the private sector Japan should build a bold developmental system.'[45] Kobayashi Hisao, a section head of AIST's Technology Promotion Office, realized the broader significance of the NGBT project to Japan as a focused policy package towards biotechnology:

> The difference between America and Japan as developed countries with a genetic engineering industry is about the same as the technological difference between us with the rocket. Since Japan is presently pursuing policy in 'fits and starts', it is amounting to nothing. We must have at once a joining of government and private sector forces.[46]

The press coverage that the NGBT received was wide; an illustration of government–media positive imaging. Japanese firms not associated with the NGBT consortium were so impressed that they initiated almost identical R&D topics in their in-house R&D programmes. A glimpse at the R&D activities of companies listed as members of the Japan Industrial Fermentation Association reveals that the majority of the 170 companies are involved in the three research areas carried out in the NGBT programme – bioreactors, rDNA or large-scale cell culturing.[47] The NGBT scheme, which MITI created to implement R&D specified in its 1980 *Vision*, had given a strong impetus to Japan's biotechnology efforts.

Planning and policy formulation in biotechnology

MITI's efforts in promoting a sense of unity and a general consistency of government policies in biotechnology precipitated a mushrooming effect among the political elite of *Nebukai*. Biotechnology was described as 'a magic cane' with so much innate potential that 'even the cat and the ladle' (everybody) rushed to be involved.[48] The STA, the CST, the MAFF, the MOE and the MHW all began in earnest to provide measures for biotechnology development in industries under their jurisdiction. Saxonhouse agrees that, 'MITI's bureaucratic entry into high visibility strategic planning for biotechnology signalled the beginning of spirited jockeying for influence among a wide array of government entities.'[49] This race among bureaucratic offices was in marked contrast to the promotion of high priority areas in the past, such as electronics and computers, where MITI totally dominated development. Mobilizing young science-educated officials, each ministry involved with biotechnology began to engage in the familiar patterns of bureaucratic sectionalism inspiring dramatic press headlines such as 'Government Bureaux and Agencies in Full-Scale Battles Over Biotech'.[50] At all levels in the central government, officials rallied to chalk up impressive ministerial achievements; the sense of urgency to expand programmes and increase budgets was overwhelming.

Activities of other ministries

There were many important precedents in policy initiatives introduced by MITI and adopted later by other ministries in their promotion of biotechnology. One of these measures was the persistent effort to establish a department office or section specifically for biotechnology promotion. The STA underwent selective restructuring of bureaux and sections twice in three years to balance interests in biotechnology. As the life sciences came to be perceived as important, so also was the demand for a separate locale other than the office at RIKEN. On 22 June 1984, the Diet formally approved the formation of the Life Sciences Planning Division in the Planning Bureau of the STA.[51] The importance of this new arrangement was the close ties of the Planning Bureau to the CST, and the increased dynamism and efficiency that the Bureau would bring to the decision-making process. Then, two years later on 1 July 1986, the Life Sciences Planning Division and the Life Technology Division were amalgamated to form the Life Sciences Division. This move symbolizes again the efforts to maintain a streamlined, efficient bureaucratic mechanism to the point of rendering a division obsolete if the problem of intraministerial co-ordination became too cumbersome.

The MHW, like the STA, also followed MITI's example and set up two offices in February 1983 to cope especially with the use of biotechnology in pharmaceuticals – widely regarded as the first area of biotechnological application. One office, that of 'Life Science', was established in the Minister's Secretariat of the General Affairs Division, while the Office for the Promotion of High Technology Pharmaceuticals was set up in the Pharmaceutical Affairs Bureau. For the MHW, the life sciences are very broadly defined as embodying 'the science and technology of human life, human living, and human existence.'[52] In June 1986, some of the burden for planning the promotion of general science and technology policy in the MHW was taken from the Life Science Office by establishing a *Shingikai* for Health and Welfare. This is again a response, inspired by *Nawabari Arasoi*, for such a body because of the critical roles advisory councils played in other ministries.[53]

The office for the Promotion of High Tech Pharmaceuticals is responsible for strengthening the R&D base of the relatively small and weak Japanese pharmaceutical industry. At the time of its inception, the *Kondankai* for Industrial Policy towards Pharmaceuticals, a private advisory body to the Pharmaceutical Affairs Chief, released its interim report providing the Office with useful information about biotechnology's development. The *Kondankai's* final report published in September 1983 suggested a list of pharmaceuticals thought possible to be developed by biotechnology and stressed the importance of soliciting co-operation between universities, industry and government to commercialize biotechnology-related pharmaceuticals.

The MOE's efforts towards establishing an office related to the promotion of biotechnology was centred on the activities of its Science Council. As discussed later in this chapter, the MOE initially became involved with biotechnology through rDNA technology. In addition, the MOE was the first ministry to compile guidelines for the regulation of rDNA experiments. In fact, it can be argued that it was this commitment as a regulator of rDNA that limited the MOE's role, until fairly recently, as an activator of programmes to train graduates in biotechnology related sciences.

The MAFF was the last ministry to respond to the bioboom of the early 1980s with a Biotechnology Office on 12 April 1984. The five officials in the office are responsible for the planning and overall co-ordination of biotechnology activity, the strengthening of the infrastructure for its development, the co-ordination and compilation of a suitable budget and have responsibility for collecting, managing and distributing pertinent information. The Biotechnology Office was formed by the MAFF's science *Shingikai*, the Agriculture, Forestries and Fisheries Technology Council. Since the commitee was administered entirely by bureaucrats its mid-term report and subsequent policy statements were influential, and incited a series of R&D projects towards biotechnology that finally culminated in the formation of the Biotechnology Office in 1984.

The revolutionary high technology biotechnology, which has been recently achieving incredible developments, has come to the actual development stage in pharmaceuticals, although it seems that in the long term, it has the potential to bring about large changes in food manufacturing and energy sources availability. It is because of this that from now on, it is extremely important that there is the promotion of the development and utilization of biological resources with biotechnology and the continued development of the technical revolution.[54]

In summary, it is important to reiterate that the bioboom, set in motion and to some extent engineered by MITI, was not the beginning of most ministries' flirtation with some aspect of biotechnology, but simply the first time in which a well-defined, co-ordinated strategy of many aspects of biotechnology-based industry was brought together as a high priority sector. Many ministries, including MITI, initially explored biotechnology by focusing on its central nucleus, that of rDNA technology.

rDNA policy

It is often forgotten by Western observers that MITI was not the flag–waver in the rDNA field. As Robert Fujimura, a technical advisor at the US Embassy in Tokyo, pointed out,

The first branch of the government to have a formal fund for recombinant DNA research was the MOE in 1977. The STA has had one since 1980, and MITI finally joined the field in 1981. Yet, due to the excellent public relations system of MITI, Western reports tend to consider 1981 as the year the Japanese government formally joined in biotechnology.[55]

Fujimura's use of the word 'formally' refers to the 'formal' funding process of receiving MOF approval for budgeted projects in Japan. Fujimura argues that strictly speaking if 'formal funding' is used as a criterion for assessing the timing of policy initiatives in a certain field, then MITI ranks third in its entry into biotechnology. He plays down the significance of MITI's NGBT project as the first instance of a government-sponsored project being affiliated with the word 'biotechnology', and also as the first attempt of a ministry to engage in a *de facto* nationally co-ordinated industrial policy defining a direction for the private sector to take in this new field. But, it is for these very reasons, coupled with the fact that the publicity surrounding the NGBT programme encouraged a biotechnology bonanza, that MITI must be clearly regarded as the flagwaver of Japanese biotechnology policy. The policy initiatives of other central government offices towards rDNA in the late 1970s were not co-ordinated as part of a larger biotechnology programme until after MITI had elevated the latter as a 'targeted' policy area.

Throughout the 1970s, academics especially those in the old, influential 'imperial' universities of Tokyo, Kyoto and Osaka, had deeply rooted interests in rDNA experiments. These public university scholars, with their access to the international science community, had been urging the bureaucracy for a long time either individually or as part of larger academic societies to scale-up Japan's efforts in basic rDNA research.[56] The JSC was particularly active as a body concentrating on urging the Science Council of the MOE to pursue active rDNA research. In 1975, it established in its National Committee for the Biological Sciences, a Subcommittee to Investigate Plasmid Issues. The next year, the same National Committee launched a Genetic Engineering Investigatory Committee, responsible for studying rDNA experiments. In October 1977, the JSC published, 'An Opinion About the Promotion of Research on Recombinant DNA Molecules in Japan.' Supporting the 200-odd scholars in the JSC, other academics had also been exercising collective pressure on government through about twenty biomedical societies.

In the late 1970s, the attitude of the government was not so much to promote rDNA projects as to regulate them. By 1978, the UK, France and West Germany had all taken the US lead in adopting rDNA regulatory policies similar to those of the American National Institute of Health (NIH). The Japanese guidelines for experimental rDNA research almost a direct copy of the American original, were finally compiled by the MOE on 31 March 1979. As Uchida Hisao, a Tokyo University professor and member of the MOE'S rDNA Technology Specialists Committee that advised the formulation of the guidelines, stressed, 'Once a regulatory measure has been taken by a US government agency, even if in the form of "guidelines", immense impacts upon international bureaucracy were inevitable, and Japan was no exception.'[57]

The guidelines had a considerable influence upon Japan. The concept itself of voluntary safety 'guidelines' to be adhered to without any supporting legislation was new. In fact, the MOE first referred to its guidelines as 'proclamation' before adopting the words, 'regulation' and 'guidelines' (the English word). The idea of safety committees in each research institution to review within certain limits their own rDNA activities was also introduced as well as that of a Biosafety Committee in the central government for periodic review of the guidelines themselves. The

MOE guidelines, however, were more stringent than their US counterparts, and were regarded as 'the most restrictive in the world'.[58] This is largely because Japan, unlike the US, did not have any biohazard laws for handling human pathogens.[59] This meant, as Uchida afterward admitted, that '... the rDNA guidelines in Japan were obliged to be self-sustaining or self-consistent without support from regulations on biohazards in general'.[60]

If Japan is observed through the *Nebukai* perspective, what seems extremely confusing is that a mere five months after the 31 March 1979 rDNA Experimental Guidelines were printed, Japan published an almost identical set of rDNA guidelines; a word for word replica of the first except for a few paragraphs. Before the differences between these guidelines are explored, the reasons for the publication of the second set by the Prime Minister's Office must be examined.

The final release of the MOE guidelines came after much stalling which Uchida largely attributes to a lack of demand by researchers for rules regarding rDNA research.[61] Although the draft of the guidelines had been drawn up as early as 1977, it was only handed to the MOE minister in October 1978, for publication amost half a year later.[62] This slow process is surprising in view of the fact that several Japanese scientists interested in rDNA attended the famous 1975 Asimolar Conference in Asimolar, California where many of the issues to be included in the 1976 NIH guidelines were addressed. This delay is perhaps best explained by the delicacy of the rDNA issue, especially when juxtaposed against the background of the contemporary Japanese public outcry over atomic power and environmental pollution. Another factor was the time taken for the MOE draft to be reviewed, authorized and endorsed by, what Uchida calls, 'a cascade of various committees'; a practice observed time and time again in the Japanese policy-making process.[63]

The main consequence of the MOE's tardiness was that the Japanese guidelines were obsolete within 24 hours of their publication. On 1 April 1979, the American NIH guidelines underwent their first revision, loosening many previous restrictions of the 1976 version. This first revision seems to have taken the MOE by surprise. At that time, the MOE had not even contemplated incorporating a Biosafety Committee. It had no mechanism to review future evolution of the guidelines because it did not think this was necessary.[64] In fact, it was not until three years later that the MOE was able to carry out its first revision of its 1979 guidelines.

The initial concern about the MOE's guidelines did not stem from inactivity, but the problems they posed to the Japanese bureaucracy's *tatewari gyōsei* administrative style – management of industry by sectionalism. The MOE guidelines were entitled, 'Guidelines for rDNA Experiments in Universities and other Research Institutions', and were formulated with only rDNA experiments taking place in universities and the MOE-affiliated national laboratories in mind. The question arose, 'What about those research bodies that do not receive grants-in-aid funding from the MOE; for example private sector research, which accounts for about 75 per cent of Japan's R&D?'

This question prompted Prime Minister Ohira on 22 December 1978 to initiate an inquiry to the CST on the status of rDNA experimental guidelines and policies for basic research. The response, entitled 'Concerning the Basis of Policy Promoting rDNA Research', was submitted to the Prime Minister on 10 August 1979. After the US had published their review in April 1979, the CST became anxious about the rigidity of the MOE guidelines, and the lack of provisions for amendments. They feared that US firms would continue to widen the gap with Japan in rDNA technology because of the benefits to be gained from the new flexible American guidelines. Again Uchida comments, '... the (Japanese) guidelines are authoritative directions deliberatively published by the Government and difficult to change from within ...'.[65]

The CST responded in the spring of 1979 by creating a subcommittee to its Life Sciences Panel, the 'rDNA Technology Subcommittee' – the committee that eventually moulded the August 1979 rDNA guidelines. There were two main differences between the CST's guidelines and those of the MOE:[66]

1. The MOE's guidelines required that all rDNA experiments not covered by the guidelines be referred case by case to the MOE minister for approval and authorization. The CST added an amendment that made it clear that those not affiliated with the MOE funding bring their queries to the STA.

2. The MOE guidelines limited the use of vectors only to *E.coli* (See Chapter Two). The CST amended this restriction to apply to *E.coli* and *Saccharomyces cerevisiae* – a response to demands by industry and the Committee on Genetic Manipulation.

The publication of the August 1979 guidelines by the CST was a rare example of the CST aiming at a general, overall measure to regulate rDNA technology. In 1980, the STA joined the initiative with a five-year research project concerning the safety of rDNA technology, which paid particularly close attention to the investigation of biological vectors, physical containment facilities, and restrictions on large-scale culturing methods. The logic behind the STA's project was that results of the rDNA safety evaluations would be readily available when demand obliged the guidelines to be reviewed.

The CST/STA co-operation in reviewing rDNA experimental guidelines has worked well. Up to summer 1986, the CST guidelines had been revised seven times. Unfortunately, the CST/STA attempt at co-ordinating national rDNA regulation for all rDNA experiments in the public and private sector has proven too much to handle. In accordance with *Nawabari Arasoi*, MITI, the MHW and the MAFF have rallied for footholds in the rDNA regulation policy game. In June 1986 MITI published industrial guidelines for biotechnology.

In June 1986 the MHW, warning other ministries to 'leave the drugs to us', drafted process guidelines for pharmaceuticals manufactured in Japan using rDNA techniques.[67] This was done to protect their jurisdiction even when American Federal Drug Administration officials, like Frank Young, had made it clear that the US, after detailed studies, saw no real need for rDNA process guidelines since the products manufactured were no different than those borne out of conventional chemical processes and could be handled by conventional laws.[68] The MAFF has also started feasibility studies on setting up rDNA guidelines.[69] Again, Uchida offers insight,

> The problem is that there is practically no governmental agency in Japan which is mandating co-ordination between ministries as far as the rDNA guidelines are concerned ... Instead, each Ministry appears to be much more concerned about claiming by the name of rDNA technology the maximum extent of new territory under its control.[70]

The imperialist laws of territorial disputes governed behaviour.

Chapter five

Implementation of biotechnology policy – strategy for its promotion

The artificial dichotomy in the policy-making process between policy formulation and implementation is merely a social scientific distinction made in order to analyse a complex procedure in smaller, more manageable parts. Like most separations of this kind, in practice there is much difficulty in clearly defining where policy decisions end and implementation of these decisions begins. The Japanese biotechnology case is no different.

Although Japanese policy decision-making in biotechnology has been long and tedious, implementation has been at a galloping pace. In Japan and Europe, decisions to initiate policy strategies towards biotechnology occurred roughly at the same time, since they were in response to developments in the US – Genentech's stock market bonanza. MITI'S Next Generation Basic Technologies (NGBT) scheme, joint agreements between Japanese and US firms, and co-operation plans between universities and industry, all had their equivalents in Europe. Japan was slower than Europe, however, in finally achieving a consensus among the different government ministries and agencies on roles to be played in promoting biotechnology. It was not, for example, until February 1986 that the MOE formulated an equivalent of MITI's Biovision. Even though European and Japanese responses to American biotechnology advances were temporally in relative agreement, the Japanese government's full participation was slower due to the need for a broad consensus among ministries.

The process of consensus in Japan generally presupposes conflict. In the jurisdictional struggles of *Nawabari Arasoi*, for example, there is a promotion of new ideas and an enticement of young officials to engage in high-spirited scrimmages for territory, money and manpower. In different circumstances, such as with the inevitability of a national crisis, the same government officials pool resources and talent in the interests of the nation.

During these alternating times of competition and co-operation, the officials display enormous drive and energy. This energy, in addition to an elitist belief among bureaucrats in the superiority of their calling, breathes a dynamism into the bureaucratic process that is especially effective in the swift implementation of long-delayed policy decisions.

In the past, the Japanese bureaucracy, especially MITI, had many effective policy tools to provide incentives for the growth of particular industries. Yet, a brief examination of the utility of former industrial policy instruments reveals that their effectiveness has sharply decreased in recent years. The bureaucracy no longer relies on outright protectionist measures, nor is it popular to implement non-tariff barriers, namely restrictions on trade and capital transfers for new strategic industries. Biotechnology has no special legislation for special tax allowances like other high priority industries such as electronics and atomic power. The government, in its advocacy of biotechnology has entered into 'high visibility strategic planning';[1] what tools has Japan used to 'nurture' the new bioindustries in the face of real competition from abroad?

Policy tools

Quotas, tariffs and non-tariff barriers

Although Japan was once a very protectionist country, except for the agricultural sector, today there are few tariffs, quotas, or non-tariff barriers. Considering that bioindustry, still in its nascent stage, is more knowledge-oriented than product-oriented, it is no surprise that there are virtually no protectionist measures since there is little to protect.[2] One very important exception arises with pharmaceuticals. Japanese pharmaceutical companies have long-been regarded as weak by international standards, and until quite recently, larger US and European firms were restricted from the Japanese market by rigid non-tariff barriers. The Japanese procedures were time consuming, costly, and demanded data from the testing of Japanese nationals only, making marketing in that nation by Western companies very difficult. As late as 28 May 1982, the Prime Minister's Office issued a public statement to the effect that because of physiological and dietary differences in the Japanese people foreign test data for pharmaceuticals could not be accepted.[3] The MHW produced evidence, for example, to show that because of dietary habits, the average Japanese had intestines which were up to 20 per cent longer than

Westerners. This, they argued, must be taken into consideration in testing drugs, since it directly affected the rate of their absorption. In the following year, however, continual US diplomatic pressure finally produced sixteen amendments in standard and certification laws *vis-à-vis* foreign pharmaceutical firms, which has brought Japan virtually in line with other nations.

There is some evidence that international pressure on the MHW to be more clear about any approval procedures has been a blessing in disguise for domestic pharmaceutical companies. In June 1986, for example, the MHW drafted succint and clear process guidelines for pharmaceuticals manufactured in Japan using rDNA techniques. Copies of this draft have been submitted abroad to foreign nations for their review and approval in order to avoid any accusations of unfair trade practices. Since these guidelines apply equally to domestic companies, Japanese firms now have the relevant information about the MHW drug-approval policies and no longer have to operate by reading the *kaoiro* (face colour) of the MHW.[4] This not only allows larger domestic firms, with sophisticated networking inside the MHW, swift approval, but also allows smaller firms, as well as those in the food and chemical business wishing to diversify their industrial bases, chances for drug approval.

Subsidies and grants

Many outside of Japan perceive that government-sponsored research subsidies are huge. There are two kinds of government subsidies; *itakuhi* or contract-styled research, and *hojokin* or conditional loans for which repayment is necessary if the project succeeds. Government support for biotechnology in both private and public sectors in Japan mirrors the support European governments have given for biotechnology R&D, around $70 million for FY 1985 (see Table 1). The US government, however, funds upwards of $500 million for the development of biotechnology; more than twice the combined finance given by the West German, British, French and Japanese governments.[5] But the perception in the West of the Japanese government co-operating 'unfairly' in a kind of 'Japan, Inc.' organizational arrangement of subsidies and special grants, is arguably derived not from how much aid is being given but by the methods that are being used. The Japanese research associations (RAs), such as the very large-scale integrated circuit (VLSI) project, have been such brilliant successes that Westerners have complained of the government and industry plotting to create unfair nationalistic R&D consortia excluding foreign companies.

Biotechnology in Japan

Table 1 Major biotechnology related projects of the Japanese Government for fiscal year 1986

PROJECT	UNIT: BN ¥, (MN $)
MITI-sponsored projects	
R&D of Next Generation Basic Tech. Project	1.28 (6.40)
Biomass Related Research	1.31 (6.55)
Aqua-Renaissance RA (Large-Scale Project)	1.07 (5.35)
Japan Key Tech Centre (PERI)	3.80 (19.0)
STA-sponsored projects	
Special Co-ord. Funds For Prom. of Sci. & Tech.	2.20 (11.0)
Resarch Development Cooperation of Japan-ERATO	1.25 (6.25)
– Hepatitis B Vaccine	1.15 (5.75)
RIKEN-Life Science Project R&D	0.36 (1.80)
MAFF-sponsored projects	
New RA in the Food and Drink Industry	0.55 (2.75)
MHW-sponsored projects	
10 Year Anti-Cancer Programme	1.58 (7.90)
Total	14.55 (72.75)

Exchange rate: ¥200/$
Sources: *Nikkei Baioteku* 13 Jan 1986 p. 2 and Table of Major Biotechnology R&D Projects (Robert Fujimura, unpublished)

In biotechnology, however, corporations, such as Mitsubishi Chemical and Kyōwa Hakkō, are playing the biggest role in promoting the industry since only these businesses have amassed the huge corporate profits needed for the tremendous investment involved. Many MITI officials complain that these large enterprises have enough retained earnings for them to be quite independent of MITI's RAs, the government subsidies as well as grants.[6] Yoshida Motoki of Japan's Industrial Bank indicates:

> The most important point to grasp is that the powerful firms in the private sector, which are mostly responsible for actual implementation of biotechnology in Japan, are already independent of (government's) administrative programmes and have their own systems in which they can carry out R&D themselves.[7]

Yet most of the large firms involved in biotechnology, despite their newly found independence, choose to engage in 'responsive dependence' on MITI within the framework of RAs.[8]

Tax incentives and borrowing

Japanese tax policy was once a very powerful and effective instrument for activating new industries. As Okimoto suggests, 'From a political standpoint, the virtue of using taxes is that it requires no direct or visible drain on the budgets and can usually be hidden from the spotlight of public scrutiny and insulated from the usual pressures of political accountability.'[9] Tax policies, however, are not very popular today. Gone are the days, as in the 1950s, when half the cost of a new car factory could be written off as a tax break in the first year of operation.[10] Although there are no specific tax codes for biotechnology, hefty depreciation allowances for laboratory construction and equipment purchase, the results of two recent tax laws, precipitated a wave of private companies interested in establishing their own laboratories. In 1981, a handful of companies had research facilities devoted to life sciences research. At present, the number is near twenty five with about as many more scheduled to erect facilities in the near future. The tax laws have also enabled private companies to use government research laboratory facilities and their personnel at a low 'rent' price.

In the immediate postwar era, underdeveloped capital markets and a propensity for saving rather than investing by potential shareholders made borrowing from the banks an important means of obtaining finance for Japanese corporations. The few, large state incorporated banks that dominated the domestic financial system were heavily regulated by the MOF, and indirectly by MITI, and their large corporate clients, therefore became more vulnerable to state intervention. In recent years, more equity issuances and retained earnings have provided additional funding for the private enterprise reducing the state's hold.[11]

It is often argued that banks still act as signal posts to direct private sector investment. But in biotechnology, despite a 1982 survey by *Nihon Keizai Shimbun* which identified this sector as having the greatest future growth potential, since March 1984, the biotechnology industry had not received any funding from two important government financial institutions, the Japan Development Bank or the Small Business Finance Corporation.[12]

San.Gaku.Kan

A policy-instrument-by-policy-instrument approach makes past industrial policy tools look rather weak. However, despite the loss of traditional instruments of economic policy and an increased vulnerability to the fiscal constraints of a large budget deficit, the bureaucracy is still quite influential. It has adjusted to a changing international environment and has shifted from its role as a body primarily preoccupied with financing companies to being a large clearinghouse of information; a think tank conducting background research to 'pick winners' and to shape specific policy directions for private enterprises. But, the demand for creativity in biotechnology, and its very broad definition affecting many areas, cannot be handled simply with government–industrial policies. Instead, the government realizes the need to promote a large-scale industry–university–government co-operation or *San.Gaku.Kan.*

The system of co-operation between *Sangyō* (industry), *Gakukai* (the academic world of both universities and national research institutes) and *Kanryō* (bureaucracy) is not simply a rephrasing of public–private co-operation in Japan. It is, instead, a recently revived concept that emphasizes co-ordinated investment and dynamic interaction between two sectors of the *Nebukai* concept, and a third sector, the academic world. In the US, this system of *San.Gaku.Kan* has existed since the Second World War, and in recent years its benefits, such as the skills of an academic research scientist turned businessman or the innovations of federally funded basic research, are becoming clearly evident. The idea of a triumvirate of industrial–academic–governmental co-operation is not altogether foreign to Japan since it was advocated by the Japanese National R&D programme in the 1970s as a framework for promoting science and technology.[13] What is new is the expression of *San.Gaku.Kan* as a coalition of energetic businessmen, determined basic research scientists and action-oriented government officials who are eager for Japan to transform fundamental research in science into a new applied form of technology. Only through an analysis of the interrelationships between *San.Gaku.Kan* can the effectiveness of Japanese state interaction and the government's policy-making process be adequately assessed.

In Japan, it has long been a criticism that the rigid institutionalization structure makes it difficult for informal communication between researchers from *San.Gaku.Kan.* Unlike in the US, frequent letter writing, unannounced laboratory visits and extensive informal collaboration have not taken place on the same scale between universities, government research laboratories and industry. There are several points that make the strengthening of this interaction special in Japan. First, this system of co-operation is only an initial step towards a larger and more important goal of international collaboration in biotechnology research; second, the bureaucracy (government) is, for the time being, the central pivot in the system; and third, unlike successes with iron and steel, shipbuilding, cars and electronics in the past, not just MITI but four other ministries and agencies are very much involved. The benefits of *San.Gaku.Kan* co-operation to R&D activity are valued highly among Japanese ministries, and many with access to an industrial base are anxiously encouraging its growth.

The MAFF in August 1984 announced its first ever equivalent of a MITI styled RA with the ministry alone sponsoring a full project involving universities and firms in a co-operative venture. The MHW, in a similar realization of the necessity of *San.Gaku.Kan*, shifted some of its former emphasis as a regulator of rDNA pharmaceuticals to that of a promoter of R&D activity in the public and private sectors. In order to develop *San.Gaku.Kan* in pharmaceutical manufacturing the MHW in April 1985 set up the co-operative Association for Promoting High Technology Pharmaceuticals, now called the Human Science Promotion Foundation. The MOE is proud of its new *San.Gaku.Kan* Co-operative Research System and describes it as

A system created in 1983 to allow the access of both researchers as well as research funding from private industry to public universities, and joint research between the professors in these public institutions and their colleagues in the private sector on common research themes.[14]

The CST in conjunction with the STA, organizations which have traditionally limited their involvement with industry, have also started to promote *San.Gaku.Kan* schemes. In June 1986, the CST promoted a bill called 'The Promotion of Research Exchange', which the CST and STA hoped would enable administration of science and technology in a large cross-section of industry – vital to biotechnology policy – and not just in a specific subset such as agriculture or pharmaceuticals.

This sudden embracing of *San.Gaku.Kan*-styled MITI tactics in biotechnology policy illustrates another reason why MITI can be regarded as a flag waver. Although rDNA projects are collectively a blatant exception, there is a consistent trend in biotechnology of MITI initiating specific R&D systems, and the other ministries following suit. Indeed, one unnamed Environment Agency official recently remarked in a newspaper interview about his tenure at the MHW, 'In biotechnology policy, the MHW is behind MITI. When I was at the MHW, I was always urged to, "learn from MITI". But when I came here, I was finally able to plan things.'[15] The copying is less a result of *Nawabari Arasoi* as more a fear that MITI, with its close ties with industry, would advance too far ahead of the other ministries.

But, how can MITI's Bioindustries Office, for example, be an initiator in biotechnology leading the way for industry, and a think tank with only a staff of eight, excluding the office head? In MITI, it is often referred to disparagingly as '*koshotai*' or small household and as '*hanninmae no ka*' or half of a section.[16] (In terms of actual numbers, it just makes half of a section.) The hidden key to the small office's ablity to co-ordinate and gather massive quantities of information about the global situation in biotechnology is through the Bioindustries Development Centre, known as BIDEC.

BIDEC

The establishment of this organization on 1 April 1983 was perhaps one of the most successful biotechnology policy strategies of MITI. It was set up as a special subdivision of the Japan Fermentation Industry Association because MITI wanted BIDEC to be formally independent of the ministry, and because biotechnology was seen then as an extension of Japan's strong foundation in fermentation and bioprocessing. BIDEC was formally established to serve as an independently neutral organization for tying together the separate components of *San.Gaku.Kan* in order to ensure effective communication between these three sectors. Alun Anderson, a long-time *Nature* correspondent on science and technology in Japan wrote, 'It is a general weakness of Japanese research and development that co-operation between large organisations is very difficult to bring about without the setting up of an independent "overlord" body to which they are answerable.'[17]

BIDEC was designed to be an 'overlord' for biotechnology to boost the industrial potential of each company through co-operation with government and university research institutes. More specifically, however, BIDEC was given the task of establishing a data bank, an international and domestic science journal, annual symposia, and both of promoting research opportunities for young researchers and facilitating international research co-operation. The feeling in the ministry was that Japan was trailing the US in biotechnological development, and there was felt a need for organization specifically to function as a 'window for international exchange' (*Kokusai kōryū no mado guchi*). Today, this window serves the double purpose of co-operating with foreign biotechnology concerns and observing developments abroad.

The establishment of BIDEC was the end result of a decision-making process that is peculiarly Japanese. As happens so often in Japanese policy-making, in order for the best practice to be adopted an overseas study group is dispatched; in this case to assess the international competitiveness of the US and Europe in biotechnology. For three weeks in October and November 1982, a delegation of ministerial officials and biotechnology-related science specialists, headed by Tanaka Masami, travelled to the US and Europe. In 1982, when Tanaka Masami was head of MITI's Bioindustry Office, the American NBF's were enjoying worldwide attention as the main focus of an international biotechnology mania. It was immediately apparent to MITI officials like Tanaka that one of the reasons for a NBF's success was the tremendously close ties fostered between industry–university–government in the US.[18] The two professors at the University of California at San Fransisco who founded rDNA technology are pertinent examples: Herbert Boyer was vice president of Genentech, and Stanley Cohen was most certainly an advise for Cetus.[19]

In Europe also, especially in West Germany and France, co-operation between industry, government, and universities had become an accepted practice. Universities, in particular, played a major role. One manager of a large West German chemical firm explained, 'If we have a research problem we cannot solve, we turn to the universities for an answer.'[20] In France, the government had been targeting universities and national research laboratories as, *poles de developpement*.[21] From this kind of evidence, it was clear to the overseas study group that first an organization for *San.Gaku.Kan* co-operation in Japan was vital, and that international co-operation should also be carried out by this organization.[22]

In fact, plans for BIDEC had already started, but they were in their formative stages. On 4 September 1982, the *Nikkei Industrial Daily* ran an article about the first concept (*Botai*) of BIDEC.[23] The newspaper reported the establishment in the Japanese Industrial Fermentation Association of a 'Research and Investigating Committee into Biotechnology Related Systems'. The committee consisted of fifteen representatives from industry such as Kyōwa Hakkō, Mitsubishi Chemical, Sumitomo Chemical, Asahi, and Ajinomoto headed by a professor of Applied Biology at Tsukuba University. Like the final version of BIDEC, the committee's role was to strengthen Japan's international competitiveness by connecting all companies and research facilities into a cross-national network of databases and state-of-the-art electronic communications.

During the autumn 1982 fact-finding trip to Europe and the US, Tanaka and others reportedly gave assurances of Japan's willingness to co-operate in international joint government ventures in biotechnology, and promoted an impression of BIDEC as a well-established, functioning organization.[24] As the trip evolved, a consensus that BIDEC was essential to Japan developed among the small group of Japanese bureaucrats, although this consensus was not derived through the usual time-consuming processes of *Nemawashi*. It was this agreement among the MITI officials during the 1982 trip which promoted BIDEC's development over the next six months as a non-profit making independent organization of MITI. Although MITI donated an initial grant to assist in its establishment, the 0ganization exists today from funding by its 180 member companies and other organizations such as the Japan Bicycling Association which is reported to contribute as much as two-thirds of the financing.[25] MITI contributions are limited to payments for specific contract research.

BIDEC's large financial support from the Japan Bicycling Association highlights MITI's ability to circumvent the tightly MOF controlled budgetary procedure, and raise money quickly with minimal problems. The Japan Bicycling Association is a discretionary fund that gives MITI access to annual revenues earned through regulated bicycle gambling in Japan. In 1982, it supplied MITI with about ¥27 billion, or 5 per cent of the MITI budget.[26] With this funding, BIDEC was established rapidly without the need for the MOF's approval.

BIDEC has an impressive appeal to other ministries wanting to aggrandize their territorial boundaries because of its wide company diversity. Over 180 firms, including more than ten foreign enterprises, are involved from fields such as pharmacy, chemicals, textiles, consumer electronics, banks, construction and so on. BIDEC is, therefore, a very convenient clearinghouse of information for MITI with an information net cast over practically the entire Japanese industry. The wealth of information gathered, therefore, is available to MITI's Bioindustries Office at relatively little cost to the ministry itself. MITI is able to use this knowledge to formulate policies and to exercise influence over the private sector. What is even more surprising is that the information network ensnares companies, such as foreign firms, outside of MITI's traditional jurisdictional web.

The information BIDEC provides MITI has proved so useful that the MHW is trying to emulate this organization with a BIDEC of its own. On 10 April 1986, the MHW introduced the Human Science Foundation Programme made up of approximately 120 companies, sixty of which are pharmaceutical firms. The other half are divided evenly among chemical, textile and food companies. Recently firms have been refusing to join more than one BIDEC organization simultaneously due to cost. The Managing Director of BIDEC, Shimizu Makoto, argues that rather than see this senseless competition for new private firms to join biotechnology organizations, it would be more efficient for the present BIDEC to act as an umbrella body serving all ministries.[27]

San.Kan relations

Research Associations (RAs)

A favourite method used by MITI to ensure the 'best practice' in government–industry relations is through research associations or consortia (RAs). Founded in May 1961 by the Mining and Manufacturing Technological Research Co-operation Law, RAs allow private companies to co-operate and pool research efforts without breaking 'antitrust' laws. Over the quarter of a century since their inception, at least fifty-two Japanese RAs have built an international notoriety for spectacular results with minimal financial and material resources. The most successful of the RAs was the VSLI project that began with MITI backing in 1976, and consisted of five major electronic companies, Fujitsu, Hitachi, Mitsubishi, NEC, and Toshiba. After two years, four of these companies challenged the US and developed and manufactured 16K random access memory (RAM) chips at prices well below those of US competitors. After a further two years the Japanese had captured 40 per cent of the world market in these chips.

In March 1980 when the VLSI consortium was disbanded, the RA had developed and marketed 64K RAM chips that subsequently captured 70 per cent of world markets. The RA also allowed Japanese semiconductor manufacturers a head start in the development of the first 256K bit VLSI circuit. Westerners were so terrified by the VLSI project results that they blacklisted the scheme as 'a computing Apocalypse' and 'a technological *Mein Kampf*'.[28]

RAs cause a problem in translation, since they are literally referred to in Japanese as research unions (of companies) – *Kenkyū Kumiai* – although the terms suggest a 'clubbing' together of companies similar to 'pre-competitive research' schemes in the UK. In fact, the original idea of 'unions' of companies for promoting industry, ironically, originated in the UK.[29] In Britain, research associations were established in 1917 by the Department of Scientific and Industrial Research to meet the technological needs of British Industry after the First World War, and have evolved as 'listening posts' and as organizations that perform 'background' research. The British government has sponsored RAs for research into non-competitive, low-priority technology areas. In 1960, a survey of industrial research in manufacturing industry found that in the UK, the more science-based an industry, the less important RAs were likely to be.[30] About the time of this survey the Japanese adopted RAs, and used them as a means of 'targeting' high priority sector R&D. The Japanese also structurally altered their version from the British archetype by disallowing co-operative research under one roof – researchers participated in the project from the privacy of their own company laboratory.

As cartels, RAs are effective in protecting certain 'core firms' (*chukaku kigyō*) from excessive competition familiar among Japanese firms; a guard against the kind of situation in the 1950s with the development of the penicillin industry. MITI organizes a handful of giant companies into 'a circle of compensation'; an arrangement that guarantees that each company is nurtured and protected by administrative guidance policy so that no single company benefits or loses in isolation.[31] The MITI officials approach 'core firms' by singling out the best in their industrial sector. This is done in order to increase the public appeal of the project, to consult with 'the experts' to make the RA successful, to tap into these firms' 'know how', and to entice other companies to join the RA.[32] This system of recruiting the most excellent and successful enterprises is referred to as 'the consignment system'.[33]

Almost invariably, firms not in RAs formulate identical R&D portfolios as those companies within an RA, resulting in the peculiarly Japanese phenomenon of myriad firms developing the same product. This reproduction of RA themes involves researchers nationwide in similar R&D engaging in vigorous competition. A partial explanation for the phenomenon is that the government, by targeting high priority areas, drastically reduces the risks of bankruptcy in that area. The central government supports this practice because it streamlines resource allocation, and leads to higher quality products; factors, they argue, justify charges of a short-term waste in national resources due to duplication of effort.

When RAs were first introduced in Japan, a joint steering committee administered them, before MITI took over the reins. The quantity of seed money proposed today by MITI for a particular project depends on three factors: the risks of the R&D, the gestation period before marketing of the product, and the licensing agreements for patents. Biotechnology R&D, by its very nature, is basic research oriented, risky, with long lead times to practical application. Any RA sponsoring biotechnology R&D, therefore, puts large financial demands on MITI.

In biotechnology, the most celebrated RA has been the NGBT project (also see Chapter 4). It is sponsored by a 100 per cent *Itakuhi* subsidy granted by MITI. Despite the project's tremendously high profile as the nation's premier project in biotechnology, it has recently suffered many serious financial and technical difficulties which cast doubt on its efficacy. First, the RA was financed using MOF annually approved grants (known as *ippan* or 'general finances'). At its onset in 1981, about ¥26 billion was promised for a ten year period.[34] By the beginning of fiscal 1986, the project was in serious financial difficulty. Only ¥6.6 billion had been given of the original budget promised which included a 1984–85 funding increase over the previous year of only ¥51 million and a 1985–86 increase of a mere ¥34 million.[35] The problem stemmed from the rising national budget deficit that had reached $500 billion, and made the imposition of low budgetary ceilings on most R&D schemes essential.

Spending cuts dug so deeply into the treasury's coffers that new biotechnology-related themes in the NGBT which were due for adoption in 1986 were dropped until the following year, and the existing three themes were stalled by an almost zero-ceiling fixed budget. A MITI official admitted that not only did NGBT's financial difficulties cause concern about its viability, but also threw the traditional practice of using large-scale RAs by the ministry into much confusion, 'Already we can say that the era of the NGBT project and that of large-scale projects is over. As long as there is a minus ceiling on the general budget, we can't achieve our objectives.'[36] This financial failing of the NGBT programme is one reason why MITI has strived to fund important organizations and more recent R&D projects such as BIDEC and the New Key Centre respectively with finance from outside the budgetary process system using discretionary funds like the Japan Bicycling Association and dividends from the privatization of public corporations.

Second, the NGBT project was also haunted by the successes of the VLSI scheme before it. Having lost control over one thousand odd patents from the VLSI scheme, MITI vowed never again to relinquish patent rights to a RA programme. The argument was that MITI would appropriate all patents generated and make them readily available on a non-discriminatory basis to non-participating firms at a small patent licence fee. However, MITI's refusal to allow participating firms proprietary rights to the patents produced, curbed incentives for technological advances. Companies sent less-qualified scientists to carry out R&D, and after the first six years, despite the numbers of patent applications, there has been little state-of-the-art technology generated.

Third, in contrast to the *Nebukai* argument there have been very strained relations between government and several firms in relation to participation in the NGBT. Although competition to join a Japanese research consortium is harsh – consisting of tough interview programmes, company presentations and so on – leading non-chemical firms such as Suntory, which had built in the previous year before the NGBT's initiation one of the first private biomedical research institutions in Japan, complained of biased selection practices in favour of firms in the chemical industry.[37] Tanabe Pharmaceuticals, Japan's largest pharmaceutical firm, flatly refused to join the RA on account of the government's decision to keep all patents produced; a move Tanabe saw as usurping its sophisticated fermentation (bioprocessing) technology in the name of updating national industry. Furthermore, Kyōwa Hakkō, the Japanese firm with the most advanced techniques in rDNA, was conspicuously absent from the team investigating rDNA. Instead the firm opted to participate in the mass cell culture research.

Edward A. Feigenbaum and Pamela McCorduck's research on the Fifth Generation Computer Project shows that these attitudes are not only limited to biotechnology.

> Resentment and hostility are hardly strong enough to describe the attitudes of another firm's managers toward the Fifth Generation. They told us frankly that they had not wanted to participate and only under duress (whose nature we couldn't ascertain) did they finally contribute their researchers to ICOT (project).[38]

Despite the slow demise of RAs in MITI, other ministries have frantically tried to promote RAs to consolidate their often inadequate relations with industry. The MAFF, for example, has been particularly adamant about stimulating biotechnology R&D in the agricultural industry. In April 1983, a Plant Nurturing Research Committee was established to review the possibility of *San.Gaku.Kan* co-operation between twelve companies, mainly from the chemical and food industries.[39] In August 1984, fourteen private companies engaged in an eight-year RA programme with the MAFF footing most of the ¥30 billion bill. Twenty-five senior researchers and their staff from four top public universities were also involved. This project marked the first instance of the MAFF initiating a MITI-styled RA venture.

The New Key Technology Promotion Centre

With the NGBT project badly floundering, there was a sense of urgency in MITI that something had to be done to help co-ordinate leading biotechnology concerns. This was to be carried out with the government providing as little as possible in the form of risk capital for basic research. This meant that the enormous financial strength of the private sector was to be used. As Tanaka Masami writes, 'Those who shoulder the development of bio-industry are, naturally, private firms. The expansion and prosperity of the industry will depend upon the extent of their activities.'[40] The New Key Technology Promotion Centre (The New Key Tech Centre) was set up on 1 October 1985 as a first step in reaching these goals. The *raison d'être* of the Centre is based on a similar philosophy as the NGBT project; to promote the three high tech areas, biotechnology, microelectronics, and new materials, proposed for development in the *1980 MITI Vision*. The Centre stresses 'newness' (*shinkisei*) in basic and applied research; the development aspects of research are of little concern. Most of the MITI and MPT officials, who organized the Centre's structure and work as its administrators, are concerned that Japan should reverse international accusations of being a borrower not an innovator of technology. They also view the Key Technology Centre as a small, yet important, influence in diverting some heat from a large trade friction problem with the US and Europe.

The centre provides a comfortable haven for businesses to promote basic research by minimizing risks and adding incentives for basic and applied research. In establishing the Centre, officials assumed that the most effective basic research should be carried out by more than one company sharing the risks and ideas. Co-operation among companies meant more brains on a project. The attitude was one of 'the more, the merrier' rather than 'too many cooks spoil the broth'.[41]

Perhaps the most remarkable feature of the Centre is that it is the result of MITI and MPT co-operation after a bitter conflict over which ministry had 'rights' for its development. Around August 1984, both MITI and MPT drew up plans for an organization to develop high technologies, especially computers and telecommunication systems. The territorial dispute between the two organizations was familiar. MPT argued that any high tech development of computers related to telecommunications was its domain; MITI countered that not only were computers traditionally its territory, but also industrialization and marketing of any telecommunications system was its business. In December of that year, MITI's 'Industrial Centre' and MPT's 'Telecommunications Promotion Organisation' were united in the MOF's first budget draft as one entity. This is vaguely reminiscent of a 'leaderless' or 'blind' policy of *Nawabari Arasoi* discussed in Chapter One.

The collaborative structure of the Centre is unique both in new methods of financing high technology-based industry, and in the strategies for basic research itself in Japan. Two or more large companies channel money into a newly formed joint-stock, subsidiary research company (*shidai kenkyū kaisha*). Seventy per cent of the finance is from public funds and 30 per cent from the private sector. The government money is not derived from the general budget, but is mainly from special, discretionary funds: the dividends of both Nihon Telephone and Telegraph (NTT) and the Japan Tobacco Inc. Also, a small proportion of finance is provided by loans from the Japan Development Bank and the Japan Import/Export Bank. The individual R&D companies established are allowed to keep all patents generated, and if unsuccessful as enterprises, the parent firms involved lose less than 15 per cent of the total capital invested. The chairman of the Centre is Inayama Yoshihiro, the Co-Honorary Chairman of *Keidanren*.

Projects sponsored by the Centre are diverse. The important difference from R&D conducted in RAs is that the Centre sponsors R&D carried out jointly under one roof, not in separate facilities. The largest biotechnology venture, the Protein Engineering Research Institute (PERI), is a case in point. As seen before, five 'core' companies were initially chosen, Mitsubishi Chemical Ltd (the representative company), Kyōwa Hakkō, Takeda Chemical, Toa Nenryō, and Toray Industries. After each of the five prompters invested 6 per cent of the 300 million yen capital for incorporation of PERI, they were joined by a further nine firms, two of which are foreign. The announcement of these two foreign enterprises, Nihon Digital Equipment Corporation and Nippon Roche Company as partisans, was a deliberate effort by Japanese officials to diffuse foreign criticism of the Centre as yet another nationalistic venture aimed at closer government–industry ties for unfair advantages in world trade.

The fourteen firms officially launched PERI on 23 April 1986 to study the structure and function of proteins and their application to industrial processes. The planned budget for ten years is ¥17 billion about three-quarters of what was promised to the ten year NGBT project before it ran into financial problems. Since the PERI project does not rely on the MOF-approved general budget, and firms are allowed to keep proprietary rights to patents, MITI officials are optimistic that this project will generate important state-of-the-art technology.[42] PERI conducts all of its R&D under one roof and ultimately the entire project will be housed in a new research laboratory in Suita City near Osaka. The original concept was for the laboratory to be closely associated with Osaka University, with all research departments to be overseen by a tenured professor from the University.

The MITI/MPT co-operative venture immediately provoked repercussions throughout central government. The *Nawabari Arasoi* reaction of other ministries involved in promoting biotechnology was to form partnerships also. The MAFF and the MOF started negotiations about a possible Key Technology Centre equivalent for agricultural development in biotechnology; the MHW and the MOF discussed the same proposal for pharmaceuticals.[43] In October 1986, the MAFF alone opened the Biotechnology Research Advancement Institution (BRAIN), analogous to the Key Technology Centre, but exclusively concerned with agricultural R&D in biotechnology. BRAIN is modelled almost exactly on the MITI/MPT Centre, right down to the statutory measures taken for its establishment. The MAFF began seriously considering the scheme in February 1985, and by September of the same year, had submitted a budgetary proposal to the MOF for approval.[44] In spring 1987, the MOF also approved the MHW's plans for its own Centre and scheduled its opening in April 1987.

Gaku.Kan relations

The '*gaku*' component of fostering growth in high tech-based industries symbolizes a break with old Japanese industrial policy relations – the academic world was not a member of the *Nebukai* coalition. In biotechnology policy, the emphasis is on basic research, and the need for creative, new ideas. Although private company research in Japan is responsible for more than 75 per cent of science R&D, this R&D tends mostly towards the applied and development end of the spectrum. Biotechnology highlights the importance of the role of the academic world in Japan, its commitment to creative research, and the significance of *Gaku. Kan* relations.

Since the government views basic research as vital to the economic prosperity of the nation, it must encourage the strengthening of supporting infrastructure for this research in biotechnology. In Japan, most policy-makers regard the supportive framework for biotechnology as weak.[45] A suitable infrastructure for basic research in biotechnology consists of three resources: the availability of qualified people, large collections of genetic materials, and instrumentation, laboratories, reagents, and information.

People resources

Japan is a nation with one of the most educated populations in the world boasting high literacy rates, a large number of higher education facilities for population size, and a larger than average number of college graduates. In science education, however, the stress has been on engineering and the hard sciences at the expense of the soft sciences, such as biology. *Keidanren* estimates that in April 1984, there were approximately 105,000 Japanese researchers engaged nationally in life science-related projects compared to about four times that number in the US three years previous.[46] MITI, however, in a 1982 survey of 200 companies engaged in biotechnology-related R&D reported that these firms employed only 4,100 qualified researchers in biotechnology. The MOE quotes the number of senior Japanese researchers in public universities working on biotechnology-related topics as no more than one thousand.[47]

Regardless of the correct figures, there is a shortage of qualified Japanese molecular biologists graduating from the nation's universities. While in the 1960s and early 1970s, European universities hummed with the bustle of newly implemented molecular biology and genetics departments, Japan by 1986 had only about fifteen public universities holding simple lecture series in molecular biology. In the US there are thirty-six times the number of PhDs graduating in biology and ten times the number in chemistry than in Japan.[48]

American R&D-based NBF's would rarely conceive of employing researchers without a PhD, whereas in Japan, non-PhDs are vigorously recruited, and trained in-house. Saxonhouse speculates that this is primarily because of traditional patterns of financing eduction and training in Japan, or it may be due, as Murakami suggests, to an inadequate, archaic higher education system for science graduate students.[49] The US government has taken almost exclusive responsibility for subsidizing training outside the firm; Japan has carelessly neglected this aspect of education.

Even after a Japanese student has earned his Phd, a rigid *Kōzasei* (chair system), and poor funding make post-doctoral research or teaching or both, very difficult. In fact, the MOE and the MHW have only recently earmarked funds especially for post-doctoral researchers pursuing R&D in Japan.[50] In April 1986, a British visiting group in a ten-day tour of Japan remarked, 'It is not uncommon for professors to have 20–30 research students. Post-doctorals are rare.'[51]

The *Kōza* (chair) consists usually of one professor, one associate professor, two assistant professors and two technical staff. Each particular *Kōza* unit is responsible for research and education in a particular topic, and receives separate funding. A *Kōza* automatically obtained ¥6 million from the central government mainly for utility and library expenses. For effective biotechnology R&D, Fujimura estimates that a *Kōza* needs at least ¥15 million in total; a figure achieved only by 15 per cent of *Kōza*, even after applications are made to the MOE's grants-in-aid.[52] In addition to the scarcity of funds, Japanese professors endure large teaching loads, and under the 'democracy' of the Japanese science grant system, an awardee is only allowed one special MOE award in a lifetime. *Kōzasei* presents other problems such as a closed university departmental system in which co-operation among *Kōza* is limited, a rigid hierarchy that stifles creativity, and so on.

Since the bioboom, the MOE has spearheaded several measures aimed at improving the education situation related to the biosciences. In FY 1986, bioscience-related R&D by Kyoto and Kyūshū University scholars received the '*crème de la crème*' of the MOE's grants-in-aid awards: the special distinguished grants of ¥251 million and ¥227 million respectively.[53] The MOE grants-in-aid are divided into three categories of special grants, two categories of comprehensive research grants, a general research section, a *shorei* research section, and an experimental research category. In FY 1987, the MOE was overhauling its 'special research grant' system under pressure from criticism that the grants were too elitist. Under the current system only well-connected scientists have a chance of winning special grant funding. With the new system the major changes will give even qualified high school teachers eligibility as part of a research team.[54]

The MOE's strong advocacy of new biotechnology-related

courses and facilities has caused a swift response from academic institutions. During FY1986, at least seven universities had genetic laboratories under construction. One of these Tohoku University, committed ¥600 million for its building. This commitment of national universities is paralleled by similar construction of biotechnology research laboratories in the private sector. In addition to revamping the grants-in-aid system and other new policies, the MOE, in the past decade, has established ten institutions collectively known as the National Research Institutes for Joint Use by Universities (NRIJU), to facilitate basic research within universities. Although these institutes are under the direct sponsorship of the MOE as bodies with university status, they emphasize basic science promotion without teaching requirements. Exchanges between NRIJUs and national, municipal, or private universities are common and four of these ten institutes are devoted to studying the life sciences.

Bioresources

In May 1982, at the height of the bioboom period, LDP politicians took the opportunity to benefit from biotechnology's 'magical cane' by initiating a 133-member Bioscience Diet Member *Kondankai*. Their representative was Kameoka Takao, a former MAFF minister, an appointment that immediately hinted at the *raison d'être* of the committee, and its close ties with agriculture. Many of the Diet members were well acquainted with agricultural issues, and excited at the possibility of biotechnology revolutionizing local farming. In December 1983, the Committee drafted its first report entitled, 'Proposal for Securing Biological Resources for the Promotion of Biosciences'. The report urged more money to be spent on the storage and preservation of biological resources by the central government.

In the LDP Committee's second report in July 1985, the emphasis was on promoting bioscience-oriented education to ready a future Japan for creative research. In their first two reports, therefore, the LDP politicians made clear statements on education and agriculture in biotechnology; issues of particular importance to their non-metropolitan constituencies.

At first glance it seems that all the central government ministries involved with biotechnology responded to the LDP incited agricultural report. In December 1983, the MAFF established a National Institute of Agrobiological Resources (NIAR) – by changing the name and organizational structure of the National Institute of Agriculture Technology – and announced plans for constructing a gene bank in the new institute. In 1984, the ministry earmarked three times its spending of the previous year for storing biological resources. In February 1983, the MHW began planning a Research Resources Bank for storing and supplying cells and genes for research, and exactly a year later in February 1984, established the Medicinal Plant Experiment Station in Tsukuba with plans for storage facilities for medicinal plant genetic resources. In April 1984, the STA, through the Resources Council – a private AB to the STA Director – initiated a 'Basic Survey on the Preservation of Potential Bioresources'. The STA also announced the need to establish a gene bank at its Tsukuba Life Science's Centre. MITI also acted in 1984 by initiating a three-year project to expand its gene bank resources and to build a Patent Microorganism Depository for storing patented microbes. In the MOE, the Bioscience Subcommittee, begun in June 1983, investigated the collecting, maintaining, categorizing, and distributing of important genetic resources as part of a plan to promote biosciences in the universities.

Although the LDP document can be argued to have precipitated this cascade of activity towards collection, storage and distribution of genetic resources, as a policy issue, closer examination reveals other considerations such as copying due to *Nawabari Arasoi*. The LDP report was a simple political authorization of policies that were already well established in the various ministries, and on the verge of implementation in the central government. Since the MAFF is the ministry with the longest history of collecting and storing resources, and in addition, has the closest links with the LDP *Kondankai*, an understanding of policies towards genetic resources can be gained by examining this ministry.

The MAFF gene bank

In microbiology, individual researchers in Japan like their counterparts elsewhere have long been concerned with the collection of microorganisms associated with the soil. Systematic collection of indigenous cultivars for rice, wheat, barley, and beans, however, began in 1953 with the construction of one genetic resource laboratory in each of three MAFF-associated institutes: the National Institute of Agriculture Technology (now the NIAR), the National Agriculture Experiment Station, and the Tohoku Agriculture Experiment Station, respectively. The collection was developed as a result of mainly personal efforts, and was used for private research projects without wide public distribution of the collected samples.

In April 1965, as a result of the restructuring of the Technology Council's Affairs Office, the MAFF started centrally administering plant genetic resources by setting up a Communication and Co-ordination Section in the ministry. In the same year, the ministry implemented a centralized seed storage and management laboratory for long-term storage of seeds collected by regional research institutes. This is the first example in Japan of a central genetic resource centre. In 1971, the MAFF expanded its genetic resources capacity by opening The Tropical Agriculture Research Centre. This centre introduced a scheme for sending one or two research teams annually to tropical and subtropical climates in search of seed varieties. In 1975, the Technology Council's Affairs Office strengthened its commitment to the Centre's new project, until by 1984, four teams were travelling abroad per annum.

The movement to construct a gene bank in a MAFF-affiliated national institute began in September 1982 with the organization of the Utilization of Microorganisms Research Committee in the Technology Council's Affairs Office. This Committee recommended a gene bank for the MAFF as the most efficient method of searching, classifying, and storing genetic material for microorganisms. In 1983, the MAFF's Communication and Co-ordination Section began planning a preliminary project to the modern gene bank, 'A Project Concerning the Establishment of a Comprehensive System for Management and Utilization of Plant Genetic Resouces and Plant Nurturing Information'. In June 1984, a report of the STA Resource Council stressing the need for a genetic resource bank in Japan increased the MAFF's chances for funding, and in April 1985 the project became a reality.

This project, therefore, was not the result of the January 1983 LDP report, but an implementation of a policy that the MAFF had been promoting internally for many years, and found in need of reformulation and implementation.[55] A partial explanation for its implementation in 1983 was the decision of the US to announce a tightening of controls on the free access system previously given to academics in biotechnology.[56] As with microelectronics, the US refused to 'export' freely its genetic resource technology to the international science community. Japan took notice.

Instrumentation and equipment

Researchers as well as bureaucrats admit the large deficit in this aspect of biotechnology, especially with experimental biochemical reagents such as oligonucleotides (DNA fragments) and restriction enzymes (proteins for specific cutting of DNA); hardware machinery including DNA sequencers and synthesizers; and information resources like databases for DNA and proteins. In Japan, the market for all three aspects of instrumentation and equipment has been growing rapidly since 1981. Most of the experimental biochemical reagents and instruments for use in the laboratory are now manufactured domestically. Imports, however, account for approximately half the equipment in a typical Japanese laboratory.[57] Most scientists on the 'cutting edge' of basic research insist on the best quality equipment regardless of national origin which in some cases means paying four times the domestic price for imported items such as radioactive isotopes, universally acclaimed Beckman Instruments' machines, or Vega Biochemicals' DNA synthesizers.

Early identification of the weakness in Japan's lack of equipment and reagents prompted a consensus, and subsequent government policy action for rectification. As early as 1980, the STA sponsored a DNA project for developing supporting industries of genetic engineering. Under the STA funding, firms such as Seiko Electronics, Fuji Film and Toyo Soda developed DNA nucleotide sequencers. Other companies collaborated to perfect DNA synthesizers and restriction enzymes. BIDEC is also planning to house a national and international database network.

National projects

In *San.Kan* relations, the activities of national research institutes are vital in the government's promotion of biotechnology. The national research institutes play a dual role in national projects acting sometimes as support centres in RA arrangements, and as bastions of pure scientific inquiry in less commercially directed schemes. In the NGBT scheme, for instance, three MITI affiliated institutes are involved in training private sector researchers and in organizing co-operative efforts in basic research. Any government, such as the Japanese government, determined to effect creative research must expect national projects to be large scale. This means an average expenditure per scheme greater than more-applied R&D, and consequently higher basic research bills.

In biotechnology, the national research centres sponsor mostly ambitious projects that require capital-intensive equipment and specifically qualified researchers. The STA affiliated Japan Development Research Centre, for example, annually sponsors a team of scientists, all under 35 years of age, in its Exploratory

111

Biotechnology in Japan

Research of Advanced Technologies (ERATO) projects to tackle especially creative ventures in science. In biotechnology, the 1982 research team, under Professor Horikoshi began searching the globe for superbugs – uniquely adapted microbes able to survive in harsh environments that may prove useful for future industrial applications.

Gaku. San

Generally regarded as the weakest link in the *San.Gaku.Kan* collaboration scheme, relations in the past between industry and the academic world in Japan were hotbeds of emotionally charged misunderstanding, intellectual separatism, and perennial mistrust. In a 1985 private survey of 150 firms involved in biotechnology, Beggs and Fayle found that less than 10 per cent had ties with a university or government institute.[58] Several factors are responsible for Japan's appalling record in *San.Gaku* co-operation.

First, until only a few years ago, many Japanese professors harboured a Marxist bias against working jointly with industry. They viewed their research in pure science as 'above' R&D efforts for commercial application. Second, unlike in the US, where symbiotic relationships between academia and business have resulted in university professors being on the boards of companies and company directors having influential positions in universities, the dichotomy between the two sectors in Japan is too clearly delineated. It is almost unthinkable for a Japanese professor or lecturer to leave his job in a rigid hierarchical academic system for the financial and career risks of a company such as a NBF. Conversely, university departments have been traditionally closed, introspective, and highly suspicious of profit-oriented researchers intervening in their work. Third, public universities and national research centres were inaccessible to non-civil servants. Private sector researchers and foreign nationals participated only in special authorized projects, or through contract research.

Saitō Hyūga, head of the Applied Institute of Microbiology at Tokyo University, suggests that if a national research institute or university was oriented towards applied research, the perception of industry was often quite different.[59] He points to his own laboratory, and the number of industrial researchers working there. These researchers were all engaged in topics under investigation at the university having periodically dropped their company research agenda to participate. Saitō also has strong links with private sector R&D through an elite, old boy network of both former classmates and students:

Executives of these companies have come through the same universities as the young students of today, and the students enter the companies with the feeling that they too can be executives. There is this intimacy between the universities and the companies and many forms of interaction between them.[60]

Professor Saitō is first to admit, however, that in the majority of cases these contacts are at best informal, and somewhat fragile.[61]

As observed before, when policies in biotechnology are being implemented, the Japanese scientific community acknowledges a general convergence of views that *San.Gaku* co-operation in both basic and applied R&D must be strengthened. Standard arguments against close collaboration, such as a distortion of educational priorities, constraints on freedom to pursue individual research, and a lack of freedom of information exchange, have all fallen by the wayside in favour of opening new avenues of collaboration between the two sectors. Japanese industrial leaders are agreeing that funding for general university activities could provide in return early access to basic science information, confidentiality in holding back certain publications until industry begins patenting procedures, or the supply of essential research software, equipment or experimental reagents. This consensus on the benefits of *San.Gaku* has started a rush for ways of intensifying the link.

Regional opposition groups – the P4 problem

In October 1980, ironically around the beginning of the bioboom in the US, the first of the opposition groups against rDNA in Japan demonstrated outside RIKEN at Wako City. This first demonstration was organized mainly by local residents who were alerted by the RIKEN Union that the research facility was planning to build a genetic-engineering laboratory on its grounds. Leaflets passed out by the local council (*jichikai*) fuelled concern, with vague assertions that the dangers of genetic engineering were far more fearsome than those of atomic power plants. Further anxiety about a biological disaster was stirred by the concurrent airing of the movie *Revived Day (Fukkatsu no Hi)*. On 21 January 1981, the first national organization of opposition groups against genetic manipulation convened. The meeting was poorly attended attracting about thirty activists, mostly housewives disillusioned with the political overtones of the anti-nuclear power movement and looking for another 'cause' at which to direct their energies. A journalist who attended described proceedings as 'Very far from the image of an anti-genetic engineering meeting; it was rather like an extremely mundane study session'.[62]

Over a year later on 27 March 1982, more than 150 people representing 39 citizens' organizations gathered in Tokyo for a convention against genetic engineering. Their final statement was paraphrased by a biotechnology news journal:

> We want to be certain that there is a formal way for our voices to be heard, that all information is readily available, that experiments are not going on behind our backs in secret, and that local residents can live rest assured around a said facility. We also demand the termination of all loosening of rDNA experimental guidelines, and the postponement of construction on any facility capable of causing a biological disaster.[63]

The last reference to the buildings of dangerous experimental laboratories was directed specifically to support the Tanida Town citizens movement at Tsukuba, which had been protesting against RIKEN's proposed plans to erect Japan's highest physical containment facility (P4) – a specially designed laboratory for carrying out hazardous experiments with rDNA – at the new Tsukuba Science city in Ibaraki Prefecture. Tanida Town is the nearest residential area to the land earmarked for the Tsukuba Life Sciences Centre where the P4 laboratory would be housed. On 8 September 1981, the town council voted against plans for the proposed facility. The situation was complicated by the scheduling of elections for a town council president on 4 October. The incumbent president did not want to make the site a campaign issue and took the platform dealing with the advantages to the region of the proposed Expo '85. The opposing candidate, on the other hand, brought the P4 issue to the floor.

By the middle of September the anti-genetic engineering movement had reached considerable proportions, and the STA reportedly 'shocked' by the opposition set up a study session (*benkyōkai*).[64] Professor Saitō Hyūga, sent as a representative to 'teach' the local residents about genetic engineering, summarized their grievances into three main categories:

1. The residents felt it was outrageous for dangerous pathogens to be used experimentally near a town where ordinary citizens lived and worked. Farmers made up a large part of the opposition, and they were upset by rumours of land prices falling, or of the biological hazards to crops.

2. The movement was skeptical of the experts' evaluation of safety, and found their reassurances unreliable (*ate ni naranai*). Like atomic power stations, the locals expected 'things to go wrong'.
3. Some residents were concerned about the ethical issues surrounding 'playing around' (*ijiru*) with the genetic components of living organisms.[65]

Saito dismissed most fears as exaggerated, attributing them to ignorance and inaccurate reports in the press about the central organism of rDNA, non-toxigenic *E.coli*. Despite the worldwide recognition of this microbe as suitable as a vector for genetic engineering (see Chapter Two), the residents were concerned about newspaper reports when a spokesman for Japan's National Health Laboratories warned that the high concentration of *E.coli* in the Tanaki River (near Tokyo), was a potentially dangerous situation. Saito clarified that any danger was surely unrelated to the use of non-toxigenic *E.coli* in any laboratory, and that high concentrations of *E.coli* found in the Tanaki River were mere markers for other related microbial pathogens probably associated with fecal contamination of the water.

On 17 April 1982, a twenty-five member committee of the local assembly reversed the decision of 8 September and agreed to support the P4 laboratory's construction.[66] The reason given publicly for the abrupt change of face was that a safety evaluation report, supported by RIKEN, about the facility's hazards to the community had been comforting. Privately, local elites admitted that politicians had persuaded them of the importance of positive imaging publicity for Expo '85, and that disruptive demonstrations or protest would impede the flow of new businesses and finance into the area.

The actual building of the P4 laboratory was delayed for over eighteen months and the project was not finished until October 1985. The problems at Tsukuba, however, still persist (autumn 1986) with demonstrations held every Saturday. The local residents have forced compromises on many issues. At least one or two recommended individuals from the town sit as part of the P4's 'Safety Management *Iinkai* Committee', and in the event of an emergency or accident RIKEN is obliged to pay compensation to the farmers. RIKEN routinely sponsors seminars, lectures, and often fields questions concerning rDNA and safety.

Conclusion: The Triple N Synthesis

The convergence of the three perspectives of the Triple N Synthesis as a framework for a systematic understanding of the policy formation process for biotechnology may seem platitudinous. Nevertheless, it is from this standpoint that the overall structure of the interrelations between the central actors must be appraised, and coherent, general principles offered as loose guidelines. It is important to note that all three 'windows' of the Triple N Synthesis necessarily highlight the essential role of the central bureaucracy, and carefully describe, albeit from three dissimilar perspectives, the bureaucracy's interactions with other actors.

Nebukai, the first of the three perspectives offered, derives most of its analytical strength by reducing almost all evidence of biotechnology policy to a doctrine of historical evolution. As hypothesized by this framework, the empirical data about the development of Japanese biotechnology policy demonstrate a strong, working relationship between the LDP, the state, and big business.

It is also apparent, however, that biotechnology policy has developed slightly differently from past Japanese industrial policies because academics are not just consulted by the bureaucracy as specialists in disposable AB packages, but feature as integrated players in the implementation of the policy they helped to create. In fact, academics from both the universities and national research institutes have often had deep concern in biotechnology policy outcomes standing to gain or lose much in social benefits or costs. Stringent laws on patenting practices, for example, deprive scientists of the free information flow deemed so necessary for basic research. Loosening experimental guidelines on rDNA reduces overhead costs for basic research and so on. No longer are the research scientists and university professors who are debating these issues in advisory groups mere neutral, value-free observers, as hypothesized in Chapter One, but they are active participants with real vested interests in the policy process. Specialist ABs have also become common modes of information exchange in many organizations involved with biotechnology. *Keidanren* and BIDEC both hold numerous *Kondankai* study sessions to debate and discuss biotechnology-related issues with 'neutral' experts.

This new role of the academic scientist demands that we take a fresh perspective on the relationship between the *Nebukai* triad and the universities. Evidence was presented of a strong sentiment within the *Nebukai* coalition – the bureaucracy, big business, and the LDP – that Japan will only develop a solid biotechnology base if a close working relationship is fostered between individual members of the coalition and the universities. Government or industrial sponsored research teams with young, innovative researchers have been crucial to Japan's efforts to stimulate basic research in the biological sciences.

The new biotechnology, therefore, a discipline which presses the cutting-edge of university-oriented basic research, necessitates a new input of qualified, innovative lateral thinkers who are not handicapped by the constraints of quick profits or dreams of product commercialization. Simply stated, from the evidence attained from analysing biotechnology policy, we need further to differentiate the *Nebukai* coalition to include the academic world of basic researchers.

Further differentiation of the *Nebukai* perspective

The biotechnology example has demonstrated that the long-term, 'stable' *Nebukai* relationships between the LDP, big business, and the bureaucracy are changing rapidly to include the academic scientists as important players. An analysis of the data can only attribute these changes to the traditional working relationships succumbing to the growing pressures of volatile domestic and international environments.

These dynamic environments are responsible for many alterations in the way the Japanese bureaucracy pursues biotechnology policy. The evidence indicates that the range of industrial policy options has been severely reduced by international pressures. Policy tools, such as tax incentives, subsidies, tariffs and non-tariff barriers, low interest bearing loans, and other examples of financial and fiscal incentives are outdated and not implemented as frequently as they once were. Large private enterprises are no longer frail financial concerns cowering under the protective wings of MITI's economic security, but multinational powerhouses with in-house revenues and policy options that make them less reliant on the bureaucracy. The policy mindset created by the 'catch-up-to-the-West' orientation of postwar industrial policy is being reversed in biotechnology through the realization that many Japanese firms have at least achieved technological parity with foreign firms. In some cases, such as in fermentation, Japanese enterprises are leading the world in biotechnological advances. The point is that Japan has become a major global player in international economics and politics with the simultaneous loss of 'small country' privileges.[1]

Continued differentiation of the *Nebukai* relationships to include an additional player does not mean that the old relations have weakened or disintegrated. The data show that on the contrary the expansion of the *Nebukai* coalition is a process whereby the academic world is seen as enhancing the government–business network. What seems to be remarkable, especially in light of the much-publicized differences between Marxist scholars and industrial leaders, is the comfortable fitting of the university or research institution, or both, into the *Nebukai* framework. Only occasionally are there isolated symptoms of growing pains in the development of the new coalition.

This last point concerning the facility with which the academics have become adopted draws some attention to what may be a powerful underpinning cultural norm in the *Nebukai* perspective – the suppression of specific, individual or institutional attitudes if they prove controversial to the larger collective interest. This seems especially true in those cases when the larger collective is that of the nation. In the biotechnology case with university scholars, for example, there was a consensus, nurtured by the state, that the national goals of economic prosperity through biotechnology development could only be achieved by the combined efforts of the *Nebukai* players and the universities. Once the academics were absorbed into the *Nebukai* framework, they began consolidating informal and formal relations, information exchange, technological transfers, and joint research, often with the gentle nudging of state-sponsored programmes. It also seems likely that there was a clear, but tacit agreement among the former *Nebukai* triad to take special care in strengthening the weak bonds between them and the universities. The importance attached to this task implies a consensual acceptance among coalition members of the state's view that basic research in biotechnology, and hence the role of the academic scientist, is vital to future Japanese economic security.

A combination of *Nebukai* and *Nemawashi*

In analysing the empirical evidence of biotechnology policy, a regularized, normative pattern emerges of the *Nebukai* and *Nemawashi* perspectives merging in a complementary framework.[2] In order for *Nebukai* to work, there is an assumption that the coalition must be very well informed, act as a unitary decision-maker, and implement decisions with a certain unity of timing. This implies a fluent, systematic, and well-ordered communication network between the *Nebukai* actors. The *Nemawashi* perspective elucidates the critical structural 'intermediaries' that act as 'go betweens' for the members of the *Nebukai* coalition in ensuring cohesion and effective communication. In fact, it is the combination of these two perspectives that provides the Triple N Synthesis with the majority of its power as an analytical tool. The *Nawabari Arasoi* perspective, discussed later, serves more of a supportive role for the first two aspects of the Triple N Synthesis being more concerned with a policy perspective that emphasizes each ministry as a player involved in a series of inducements and sanctions that direct its behaviour.

What are some of the structural elements revealed by the *Nemawashi* perspective that serve to forge close *Nebukai* relations? In public–private sector relations, for example, the most important of the 'intermediaries' for MITI in biotechnology policy were the information-gathering organization, BIDEC; the RA alternative, the New Key Tech Centre; and the system of RAs itself. Described primarily in government literature as a think tank, BIDEC is more than a provider of vital technical resources. In addition, it listens to the demands of industry, articulates the requests of MITI, forges binding consensus among its members, and takes an arbitrator's role in co-ordinating the delicate balance between interfirm co-operation and conflict. Since over 180 firms are members of BIDEC, MITI has carefully structured a wide-ranging quasi-government organization to bridge the state–private sector gap.

Similarly, with the Key Tech Centre, its initial justification for establishment was not explicitly to cultivate closer ties with industry. In fact, the Key Tech Centre emerged from MITI's need for alternate, more flexible financing of biotechnology-based firms than the MOF would allow. As discussed in Chapter Five, the Centre only materialized after a broad agreement with the MPT following a bitter *Nawabari Arasoi* contest. The success of both BIDEC and the Key Tech Centre in contributing to the close state–private sector relations of the *Nebukai* coalition, whatever the reasons for their formation, is clearly evident from the emulation and rampant copying they have gained from other ministries.

RAs initially unique to MITI's organizational structure have proved so effective in forging close *Nebukai* relations through the years that in biotechnology policy other ministries have adopted them as standard structural features in their relations with the private sector. The MAFF, for instance, selected the opportunity of supporting agricultural projects in biotechnology as a pretence for establishing its first ever joint public–private RA in 1984.

Another perspective from which to buttress this argument of the combined nature of *Nebukai* and *Nemawashi* is to review earlier examples of close relations between *Nebukai* coalition members. Particularly, if the focus is on state–big business relations, there must have been analogous organizations to BIDEC, the New Key Tech Centre, and RAs to explain effective interpenetration across the private–public divide. In the consumer electronics field, for example, the JECC (Japan Electric Computer Corporation) was an equivalent of the Key Tech Centre since it played the critical role of providing adequate financing in assisting Japanese computer companies in the computer leasing business. In the high technology field of information technology, BIDEC is mirrored by the Information Technology Promotion Agency – a non-profit organizational umbrella for projects in this field, Again in consumer electronics, the celebrated Very Large-Scale Integrated Circuit (VLSI) research association was the forerunner of biotechnology's NGBT equivalent, and gained much notoriety in the US for its 'unfair' comparative advantages in promoting the Japanese semiconductor industry.

Close *Nebukai*-style relationships between the bureaucracy and private enterprises have also been entwined in biotechnology by various advisory groups. Although located on the fringes of the bureaucratic machinery, these groups were considered essential to the *Nemawashi* analysis. In the *Nemawashi* perspective, complex administrative problems were broken down into manageable, often repetitive tasks, and given to specialist advisory groups for reflection. This provided adequate opportunity for two-way partial interpenetration of both the private enterprise into the bureaucratic policy formation process, and the bureaucracy as an interventionist influence on individual firm R&D policy.

Similar regularized patterns of communication have emerged between the bureaucracy and the LDP as well as between the state and the academic world. In the fomer, the information channels again served effectively as two-way communication networks doubling as opinion dissemination devices to support the bureaucracy's arguments and forge a consensus among the factions of the LDP. Speeches to the LDP Biosciences committee by bureaucratic leaders resembled classroom lectures rather than analytical discussion fora.[3] Informal and formal information sharing between politicians and bureaucratic elite led to universality of attitudes and both mutual trust and confidence of respective positions.

The appointment of the late, former MAFF minister Kameoka Takao as head of the LDP Biosciences Committee clearly demonstrated the focus of the ruling party's interest in biotechnology. The *Nemawashi*-styled communication and sociopolitical ties between the former minister and the MAFF were secure enough to ensure that the latter promptly signalled a positive sign of approval to any policy statement on agriculture that the LDP committee engendered. Although, it is argued that because of the timing the MAFF response of a budgetary increase for agricultural projects concerning biotechnology came too early for it to be inspired by the LDP, and that moreover the LDP statement was perhaps even initiated at the provocation of the MAFF (evidence that proved difficult to obtain), it is clear that the LDP Biosciences Committee took political credit for prudently discussing and then compiling a report on a previously neglected policy area. For what it is worth, the MAFF was satisfied with the arrangement of the LDP politicians enjoying the limelight, and was content to partake in the spoils of an increased budget and an enhanced biotechnology programme.

Using the Triple N Synthesis as presented above, three questions concerning the formation and implementation of biotechnology in Japan are answered. In attempting responses to these questions, both the strength of a particular focus of the Triple N Synthesis as well as the combined power of the triad are freely employed. In this way three perspectives are presented, not as mere analytical kits producing more convenient explanations of certain data than others, but as component parts of an integrated, functional synthesis providing different, yet necessary views of a dynamic system.

Why does the state play such a strong role in the development of biotechnology?

The very nature of biotechnology, a technology still in its formative stages, suited it to MITI's long experience in nurturing fledgling industries. The uncertainty about the future of biotechnology also proved unattractive to the private sector which was unwilling to tackle the long-term economic risks of biotechnology development alone.

Mutual trust and confidence between coalition members has led to a clear definition of roles

The *Nebukai* perspective argues that the bureaucracy won the trust of other members of the coalition in providing an effective leadership. Through the years, the state, based on this mutual trust and confidence and the advantage of an impressive record with similar tasks, has cultivated a well-defined role in the coalition that places the ultimate responsibility of formulating biotechnology policy with it, the state. Even though the *Nebukai* perspective suggests unity of timing and action of coalition members, the effective *Nemawashi* communication ensures that each member is functionally independent. Combination of the two frameworks provides a picture of a cohesive coalition benefiting from separate inputs from each member on assumption of a final, mutual agreement. This sense of separate identity among coalition partisans allows a clear delineation of roles in the coalition, and accounts for members exercising power even in policy areas where another participant has enormous vested interests.

Indeed as hypothesized, in biotechnology policy there is a definite pattern of the LDP granting almost full autonomy to the bureaucracy, especially MITI, to formulate, administer, and implement industrial policies. Even though environment pollution in the early 1970s had proved something of a thorn in the side of the ruling party, only passing attention was given in the LDP Biosciences Committee to regulatory policies for containing rDNA 'superbugs' and for policing the potentially dangerous environmental effects of rDNA. Despite strong interests in regional policy questions, LDP members had less input than the bureaucracy in the promotion of bioindustries in rural areas. Conversely, however, the LDP balanced this power by its enhanced role in agriculture, cancer R&D, and educational policies towards biotechnology development. Noticeably, the first two major reports of the LDP Biosciences Committee were on agriculture and education, and it was the MAFF and the MOE ministries respectively that demonstrated the most response to these reports.

Also in line with the *Nebukai* and *Nemawashi* hypotheses, the opposition parties were either too weak or too disinterested to make rDNA and the environment points of contention. The exclusion of the opposition parties from the coalition and their lack of communication access into the bureaucracy made the Clean Government Party's (*Komeitō*) attempt to take a stance fade as quickly as it had begun.

Co-ordinating role of the state in managing a mixture of private sector organization and market forces

The state served initially as a central co-ordinator reallocating resources to the new bioindustries. Long-term plans, like the 1970 and 1980 MITI Visions, were possible to conceive because of the stability of the *Nebukai* coalition and the rapid communication and feedback between sectors of the *Nemawashi* perspective. It was the state's position in the centre that allowed the combination of policies towards declining industries on the one hand and expanding sectors, such as biotechnology, on the other. Just as MITI cleverly administered reductions in the excessive capacities of the shipbuilding and aluminium industries without major bankruptcies, the ministry also coordinated the adjustment process for the Japanese chemical industry by shifting to a new biotechnology basis of production.

In bureaucracy–big business relations, a combination of the *Nebukai* and *Nemawashi* perspectives is important in understanding the mixture of market forces and organizational factors employed by the state to promote biotechnology policy. Okimoto suggests that

> organizational direction is necessary in Japan because of the nature of the country's complex and finely meshed industrial system – a system that functions on the basis of competition and co-operation, market and hierarchy, public and private sector interpenetration, government/business interdependence, and consensus. The state is, in some senses, the integrative mechanism that keeps the whole system functioning as a comparatively cohesive whole.[4]

Although vigorously involved in competitions to commercialize similar biotechnology-based products, large Japanese firms have been quick to co-operate – hastily joining the NGBT project, becoming members of BIDEC, supporting the New Key Tech Centre, and regularly discussing issues in *Keidanren*'s Life Science Committee meetings. Some of these endeavours offered little to assist firms in biotechnology development. The NGBT project, for example, has minimal potential for generating state-of-the-art technology, is in constant financial difficulties, has any new technology produced appropriated as the property of the state, and has compromised for many leading private enterprises their competitive advantages. Yet the large firms still continue to participate claiming that RAs prevent wasteful duplication of R&D, pool finite resources, and generate cost-effectiveness in achieving economies of scale. Similarly, government demands for more creative, basic research were readily adhered to by individual firms that adopted more basic research oriented investment portfolios even with the knowledge, as has been seen, that there was no guarantee that more basic research would translate into the development of new products and quick profits.

Contradictions to the strong state argument

Often, the evidence gathered from analysis of the biotechnology decision-making process seems contradictory to the theory of the strong state having an effective role in co-ordination. Large corporations in biotechnology seem to be financially less dependent on the central government than before, largely through retained earnings and more equity issuances. This is coupled with the perceived loss of power suffered by the bureaucracy. MITI, in particular, no longer casually wields policy instruments such as import duties, tariffs, quotas, non-tariff barriers, restrictions on foreign direct investments, and so on in order to control domestic industry. Even among the other ministries, MITI seems to be losing political influence as demonstrated by the compromise with the MPT over the New Key Tech Centre mediated with the help of the MOF, or the difficulties in finding additional sources of biotechnology funding.

Even with the much celebrated RA system (although the bureaucracy did take an effective leadership role and exerted strong pressures in the initial establishment of the NGBT project, for example) the private sector's behäviour was not simply one of a 'follow-the-leader' mentality implied by a *Nebukai* and *Nemawashi* analysis. Tanabe Pharmaceutical's refusal to join the NGBT project because of an unwillingness to share its advanced technology is a salient illustration of non-compliance with the government. Similarly, the *Nebukai* concept of mutual confidence and trust between sectors is undercut by MITI's vigorous, yet unsuccessful efforts to lobby the Diet to approve a 'Special Temporary Development Law' for biotechnology.

Despite the apparent decline in influence of the bureaucracy, especially MITI, large firms still adhere to MITI's recommendations, and business federations such as *Keidanren* show continuous allegiance to requests made by MITI. This behaviour of the private sector demonstrates the importance of collective interests in Japanese government–business relations. *Nemawashi* provides some clues to the reasons for the private sector's lingering persistence to remain acquainted with the bureaucracy.

A new role for the state?

The role of the bureaucracy is no longer that of a financial resource to a weak private sector. In biotechnology, even though special funds such as the dividends from the shares in the privatization of NTT and the Japan Tobacco Inc., as well as alternative financing from the Japanese Bicycling Association provided additional flexible sources of revenue outside budgetary constraints of the MOF, the amount of money was too small to make any real impact. MITI's main role in biotechnology policy, and indeed that of the entire central bureaucracy, has been centred on the procurement of information, a policy direction finder, and a forger of broad consensus through *Nemawashi* channels.

The state had made it clear that it was relying heavily on industry to provide the finance and technological innovations in biotechnology, and industry responded to the direction given by the bureaucracy. With the help of the technology policies of the state, the private sector began cultivating new relations with the formerly recalcitrant academic world, R&D themes from the government-sponsored NGBT project rapidly influenced Japanese biotechnology firms, and there has been a conscious effort to be creative in research.

The persuasive power of the state has been clearly evident in the 1981–2 bioboom engineered by MITI, in which the Japanese bureaucracy effectively woke the private sector, the academic world, and the LDP out of their biotechnology slumber. The aggregation of firms into business federations such as *Keidanren* catalyses this ability to cultivate such an impact on the private sector since the bureaucracy can distribute a clear, succinct statement to one organization, and in most cases, expect a precise, unitary response. The speed of consensual decision-making is also enhanced within *Keidanren*, for example, by the further differentiation of the firms involved with biotechnology into a specialist standing committee, the Life Sciences Standing Committee. The strong collective interests of private enterprises in *Keidanren* facilitates the state's efforts in finding a balance between the need for competitive market forces and for co-operation in interfirm organization. The *Keidanren* forum and the Life Sciences Standing Committee are at least negotiating starting points for companies in harsh competition and at most co-ordinating centres from which the state can provide organizational direction.

The impressive public relations effort of the bureaucracy to create a realization of the importance of basic research to biotechnology development has led to the relatively painless adoption of the academic world into a system of co-operation with private industry. In addition to the *Nemawashi* channels, the news media have played a vital role in disseminating technical information about biotechnology to the general public.

International perceptions of Japanese biotechnology

The degree of power sharing and balancing of interests in biotechnology among coalition members in the *Nebukai* analysis and the effective communication provided by the *Nemawashi* perspective are often interpreted by outside observers, in particular, as 'Japan, Inc.' – a nationalist plot to dominate the rest of the world economically through a centralized, goal-oriented strategy. Foreign governments and the private sector abroad view 'the Japanese' targeting of biotechnology as a high priority area with much trepidation. For the Americans, especially, Japanese 'targeting' practices are 'unfair' because they enhance Japanese private sector competitiveness in world markets at the expense of US products.

Based on a lack of information about the Japanese government policy-making mechanism, paranoia, fear and minimal under-standing about the development of biotechnology are spurred by Japanese achievements in other strategic industries. 'The Japanese', in this context, is the monolithic national actor of the conservative coalition functioning as a 'centrally, co-ordinated smooth running machine'.[5] The actions of one member of the coalition, the private sector for example, with its advertisement campaigns, investment portfolios, and R&D programmes, reflect policy initiatives of the other two. Billboard advertising in Tokyo for large firms reflect policy platforms of the LDP and the ministerial officials.

Many misconceptions about Japan are conceived from outdated ideas of either Japanese industrial policy practices of the 1950s and 1960s, or the interventionist role of MITI. Perceptions in the West that focus too heavily on anachronistic policy tools or

MITI's past glory suggest that MITI's use of these tools facilitate a kind of symbiotic organizational arrangement between Japanese public and private sectors of subsidies, special grants, low interest-bearing loans and tax incentives which give the Japanese government 'superior' ability to control big business. The 'notorious' Japanese research consortia are a case in point. The very large-scale integrated circuit (VLSI) project of the late 1970s was so successful that Westerners complained of the government and industry creating unfair nationalistic R&D programmes that systematically excluded foreign companies.

Conversely, the *Nebukai* and *Nemawashi* perspectives help to explain why the Japanese government reacts to weaknesses in biotechnology's infrastructure as symptoms of immaturity in Japan's development *vis-à-vis* the West. This argument suggests that the LDP, the civil service, and the private sector perceive themselves as a monolithic unit in a desperate situation – both inferior to their functional equivalents in Western nations and under constant pressure from the Japanese people to maintain high international standards at home. As the frequency and intensity of the perceived threats increase, so does the desire of the coalition to take action. Paths of specific policy action under these constraints are often not the result of normative judgements and engender irrational behaviour. The intensity of diplomatic pressure from the US on Japan to help reverse the US trade deficit to Japan, for example, can be thought to have precipitated excessively ambitious and altruistic international policy initiatives in biotechnology. The offer of a ¥3.3 billion Human Science Programme to rid man of all major diseases, originally proposed as a non-military alternative to the American Star Wars – Strategic Defence Initiative – is perhaps best seen as a desperate attempt by the Japanese to convince the Americans of their commitment to broad international problems.

Many bureaucratic elites are guilty of deliberately overstating Japan's disadvantages to the West in biotechnology guideline reports in order to motivate an intense feeling of nationalism, and thus incite the *Nebukai* coalition to action. This is often described in the media as the 'Japanese complex towards the West'. In discussing the development of a suitable infrastructure for genetic resources, for example, a MITI committee stresses that:

> In the United States and in various European countries, there are plans for the systematic preservation of biological resources, and from a future prospect of bioindustries, in the modern

Japanese system there is much backwardness, and in its turn, has the potential to induce a barrier to healthy development of our nation's bioindustries.[6]

The word, *Ōbei*, meaning both Europe and the US, or more effectively 'the West', is often used in statements such as these even if, as is the case in biotechnology, it is only the US and not Europe who is ahead. In addition, *Ōbei* is frequently placed in contrast to 'We the Japanese' (*Wareware no Nihonjin*); words that are bound to accentuate the already strong nationalist interests of other *Nebukai* coalition members.

The supportive role of the *Nawabari Arasoi* model

The *Nawabari Arasoi* perspective acts as a supplementary theory to *Nebukai* and *Nemawashi* in the Triple N Synthesis. It is especially useful to explain turmoil during budgetary conflicts. These bureaucratic battles are often wasteful, producing duplication of similar policies which target the same objects (such as the Key Tech Centre), or are uncoordinated such as the regional policy programmes of MITI, the MAFF, and the MPT. It is a testimony to the strength of the felt national collective interest that the acrimonious debates are quickly forgotten once a compromise between the ministries is reached. The *Nawabari Arasoi* framework is not limited only to budgetary disagreements, but also provides in-depth analysis of other problems of biotechnology policy.

The framework, as a supportive perspective, offers alternate reasons for MITI's involvement in biotechnology policy. MITI's engagement in biotechnology stemmed, it is argued, from a need for a major new pro-growth programme that would be highly visible and welcomed by industry. The subsequent idea of combining policies for declining industries with biotechnology is interpreted as MITI's attempts to maintain and enhance bureaucratic prestige and influence. In biotechnology policy, copying among ministries is a very persistent obsession, so much so that three out of the five main ministerial actors have their own cancer programmes completely separate from each other with little collaboration.

Since biotechnology has been heralded by the Japanese press as 'the last technological breakthrough of our century', and given tremendous expectation as a means to revive declining industries, most ministries are inducing private companies under their jurisdiction to diversify their interests into biotechnologically produced goods. Thus, former food companies, like Ajinomoto, that have been traditionally under the jurisdiction of the MAFF are suddenly producing pharmaceuticals and finding themselves under the wrath of the MHW. Moreover, the race to promote new substances has given rise to biotechnological products that can be classified as either food or drugs, or as needing regulation by-product or process criteria.[7] Since duplication of effort and regulation can be redundant in cases where anachronistic rules apply to new products, careful steps must be taken to ensure that *Nawabari Arasoi* does not result in foolish compromises among bureaucratic actors.

The *Nawabari Arasoi* perspective also offers another view on the frequent Japanese reticence and inaction in choosing among difficult options. In this view, the options are not deliberately picked, but are delayed resultant actions of intense competition between ministries. In the effort to balance interests, ministries fall back on traditional tactics of conflict resolution:[8]

(i) avoiding controversial issues by delaying decisions on them, and referring these decisions to other bodies for resolution
(ii) compromise and logrolling, that is, trading of subordinate interests for major interests
(iii) expressing policies in vague generalities, representing the lowest common denominator of agreement in which all can acquiesce
(iv) basing policies upon assumptions which may or may not be realistic

The resulting immobilization in both the decision and implementation of policy, although often attributed to a wait-and-see policy directed towards the US, may be due also to *Nawabari Arasoi*.

Which came first: push of the state or pull of industry?

In the study of postwar Japanese industrial policy, the question is often asked which sector is responsible for industrial development – public or private? In attempting a response, the advantages of using the Triple N Synthesis rather than the 'state-led capitalism' versus 'market forces' approach is that the former avoids focusing too much attention on an analysis of the two sectors as distinct entities and considers them rather as part of a whole system coordinated by the state. In the case of biotechnology, the question of which came first, the state or the firm, involves a complex mix of state and private sector interaction.

Despite the identification of biotechnology by MITI as early as 1979 as an important area of future economic growth, and the subsequent publication in March 1980 of *MITI's Vision of the 1980s*, it was not until the autumn of that year that MITI really began a feverish campaign to boost this technology in Japan. This was a direct result of provocation from the domestic private sector after the explosive sale of biotechnology equity in the US. Large Japanese private enterprises, communicating mainly as a collective through the business federation, *Keidanren*, pressed MITI to develop fully its half-baked policy initiatives towards biotechnology or witness Japanese industry fall technologically years behind the Americans.

The means for the initial identification of biotechnology in the *1980 MITI Vision* as a potential growth industry of the 1980s fits a classic information procurement model or *Nemawashi* analysis. The state's effective process of consultation through its wide, branch-like communication network provided important clues for the direction of future policy. But before these networks could be used to forge a broad consensual agreement over a wide political spectrum about the importance of biotechnology policy, events in the US did the consensus building job for the government, at least where the private sector was concerned, and caused a tremendous demand for a state-led coordinated biotechnology policy initiative.

The state responded with the NGBT project and a massive public relations campaign attesting to the RA's importance. At the same time, the bureaucracy was successful in creating a consensual awareness among the key players of the *Nebukai* coalition about the necessity of basic research and creativity to the development of biotechnology. Programmes were launched, attitudes towards academic scientists reversed, and new organizations built to enforce the change.

In summary, although the private sector can claim responsibility for initially arousing Japan's awareness of the importance of biotechnology, it was the state who nurtured and directed the development of biotechnology until the private sector appreciated its value as a profitable enterprise.

How effective is Japanese biotechnology policy?

How do the biotechnology data affect the question asked in the introduction about the relevance of past Japanese policy practices to biotechnology's present and future role in the Japanese economy? Does the differentiation of the traditionally close *Nebukai* interrelationships to include the academic scientist distort the argument made by foreign competitors that since Japanese-styled industrial policy worked before, it can work again?

It is easier to field questions about the strengths of Japanese competitiveness and the possible recurrence of their impressive past performance by interpreting views about Japan held by outside observers. Most mass media accounts and even foreign government reports err by attaching too much importance to the effectiveness of specific structural features of the policy-making system, such as the strength of biotechnology policy tools to support the private sector financially. Indeed, even the celebrated 1984 OTA report on Japanese biotechnology, which emphasized the strength of the NGBT project as that nation's secret force to eclipse world markets, has proven inaccurate. The evidence of biotechnology policy demonstrates that the NGBT project is widely thought to have already failed; an RA victimized by the MOF's tighter controls on the budget. Even reports that focus on MITI's midwifery role for the private and academic sectors miss the point, and portray MITI as it was over two decades ago.

The problems of little venture capital financing for innovative NBFs, and the difficulties inherent in a policy engineered by the state to engender scientific creativity complicate assessing the effectiveness of the state as a political institution, or evaluating the policy tools it wields. These problems persist along with fragile academic–business relations, and a lack of an educational starting ground for pure basic researchers in the life sciences. The only safe assumption to make about the effectiveness of Japanese biotechnology policy is that any analysis must start from examining the relationships between the actors involved with this policy.

The critical synthesis of the Triple N Concept ensures that these relationships remain dynamic, flexible and close. This includes the continued importance of the co-ordinating power of the state, mutually beneficial working relationships between biotechnology policy actors, a deeply rooted government–business interdependence, effective two-way communication channels, 'intermediaries' in linking various sectors, and the means of accessing information necessary to resolve disputes among competing ministries. In practical terms, this means being flexible in finding alternate sources of financing, promoting consensual agreements when national interests are concerned, preparing for the possibility of political power shifting from MITI to other ministries, and so on. Although it is too early to speculate about specific changes, any dramatic alterations in the system, such as the shifting of the balance of power between ministries or the liberalization of financial markets for increased capital would only be effected under a collective agreement that the interests of the nation are not in jeopardy.

The final synthesis of the three perspectives is presented as an example of 'typological' theorizing – a fitting of the Triple N Framework to the biotechnology environment. In short, it seems difficult to assert whether or not the general principle of dynamic interaction between the public and private sector in explaining policy formation, which works so well in this particular context, would be viable in abstraction from biotechnology.

Indeed, the evidence found in an analysis of the biotechnology situation often seems arbitrary or paradoxical. This includes: vigorously competing firms co-operating for joint research, close state supervision of developmental policies juxtaposed with a healthy display of market forces (the notorious market versus organizational dilemma), ministries in bitter conflict and then suddenly uniting, and Marxist academics, embracing big business in co-operative research. Whether or not a similar synthesis of *Nebukai*, *Nemawashi*, and *Nawabari Arasoi* can organize different sets of data for other industrial sectors in Japan is left for others to assess.

Notes

Introduction

1 Michael Crozier (1964) *The Bureaucratic Phenomenon*, London: Tavistock Publications, p. 196.
2 Samejima Hirotoshi interview, 24 Sept. 1986.
3 BIDEC Member Company Directory, Special Issue of *Japan Bioindustry Letters*, Tokyo: BIDEC, April 1986.

Chapter 1 Understanding Japanese policy-making with the 'Triple N Synthesis'

1 James Abegglen (1970) The economic growth of Japan, *Scientific American*, 222; 31–7.
2 US International Trade Commission (1983) Foreign Industrial Targeting and its Effects on US Industries, Phase I: Japan. Washington DC: US Government Printing Office, p. 20.
3 Peter Daly (1985) *The Biotechnology Business*, London: Frances Pinter (Publishers) Ltd, p. 60.
4 G. C. Allen (1981) *The Japanese Economy*, London: Weidenfeld and Nicholson, p. 107.
5 J. A. A. Stockwin (1982) *Japan: Divided Politics in a Growth Economy*, 2nd edn. London: W. W. Norton, p. 154.
6 *Commercial Biotechnology: An International Analysis*, Washington DC: US Congress, Office of Technology Assessment, Jan. 1984, p. 21.
7 George Ruthmann (1985) 'Research, Commercialization and the Future' in Sandra Panem (ed.) *Biotechnology: Implications for Public Policy*, Washington DC: The Brookings Institute, p. 48.
8 Albert Gore (1985) 'Congressional Perspectives' in Panem, p. 18.
9 Ronald Cape (1986) 'Future Prospects in Biotechnology: A Challenge to United States Leadership' in Joseph G. Perpich (ed.) *Biotechnology in Society: Private Initiatives and Public Oversight*, New York: Pergamon Press, p. 9.
10 Yamamura Kōzō (1986) 'Joint Research and Antitrust: Japanese vs American Strategies' in Hugh Patrick (ed.) *Japan's High Technology Industries*, Seattle: University of Washington Press, pp. 171–211.

Notes

11 *Baioindasutorī Shinkō Iinkai Hōkokusho (Report of the Bioindustry Advisory Council)*, Tokyo: MITI, Aug. 1984.

12 *21 Seiko o Hiraku Baioindasutori – Sono Tenpo to Kadai – (Opening up the 21st Century with Bioindustry – Its Prospects and Problems –)* Tokyo: MITI's Basic Industries Section Press, 1984, p. 95. (Parentheses are mine.)

13 Ronald Dore (1983) *A Case Study of Technology Forecasting in Japan: The Next Generation Base Technologies Development Programme.* London: The Technical Change Centre, p. 5.

14 Chalmers Johnson (1982) *MITI and the Japanese Miracle*, Stanford: Stanford University Press, pp. 275–90.

15 *White Paper on Science and Technology* (Summary) (1984) Tokyo: Science and Technology Agency, pp. 13–14.

16 James E. Anderson (1975) *Public Policy-Making.* New York: Praeger, pp. 9–10.

17 Margaret A. McKean (1977) 'Pollution and policy-making' in T. J. Pempel (ed.) *Policy-making in Contemporary Japan.* London: Cornell University Press, pp. 201–39.

18 Daniel Okimoto (1986) 'Regime characteristics of Japanese industrial policy' in Hugh Patrick (ed.) *Japan's High Technology Industries.* Seattle: University of Washington Press, p. 59.

19 ibid. p. 90.

20 J. A. A. Stockwin (1986) Lecture Nissan Institute, Oxford, 7 Feb.

21 Graham Allison (1971) *Essence of Decision: Explaining the Cuban Missile Crisis.* Boston: Little, Brown, p. 67.

22 ibid. p. 76.

23 ibid. p. 68.

24 James March and Herbert Simon (1958) *Organisations.* New York: Wiley.

25 ibid. p. 177.

26 Graham Allison (1971), p. 72.

27 Michael Crozier (1964) *The Bureaucratic Phenomenon.* London: Tavistock Publications, pp. 175–6.

28 ibid.

29 Morishita Noboru (1985) Shingikai Ni 'Shingi Ranpu' Tentō (Lighting a 'Deliberative Lamp' on Deliberation Councils), *Tsūsan Jyanaru (MITI Journal)*, November, p. 46.

30 Professor Murakami Yoichirō interview, 12 Sept. 1986.

31 Ehuh Harari *The Institutionalization of Policy Consultation* (unpublished), p. 1.

32 Sone Yasunori (1986) Yarese no Seiji: Shingikai Hoshiki o Kenshō suru, in *Chūō Kōron*, January, in ibid.

33 Professor Murakami Yoichirō interview.

34 Morishita Noboru interview, 20 Oct. 1986. Also Morishita (1985) pp. 46–7.

35 *Nature*, 305, 361.

36 Morishita interview and Morishita (1985) p. 48.

37 *Nihon Kingendaishi Jiten*, Tokyo: Tōyō Keizai Shinposha, 1978, p. 448.

134

38 Morishita (1985) pp. 46–7. Chalmers Johnson's account of the development of the present day nine electric companies is unclear. He asserts, without references, to the 1938 Electricity Power Control Law which set up only nine companies (with no distinction between generating (*hassō*) and distribution (*haiden*) companies) and overlooks completely the fact that these were only distributing companies for the large Japan Electric Power Generating Company of which there is no mention. The nine companies, as Johnson depicts them functioning as both generators and distributors of electricity, did not develop until 1951, the direct result of the dissolution of the Japan Electric Power Generating Company and the proposals set forth by the aforementioned *Shingikai*. (See also Morishita, pp. 47–8; *The Companion to Japan's Industry (Nihon Sangyō Dokuhon)* and *Nihon Kingendaishi Jiten*, p. 448.)

39 Morishita Noboru interview, 30 Oct. 1986.

40 Morishita (1985) p. 47.

41 Miyata Mitsuru interview, 31 Oct. 1986.

42 Harari (unpublished).

43 Allison (1971) p. 154.

44 ibid. p. 169.

45 *Baio Gyōsei o Ōu (Following Biotechnology Administration)* Tokyo: Nihon Kōgyō Shimbun, 1986, p. 14.

46 Allison (1971) pp. 144–82.

47 ibid.

48 ibid. p. 154.

49 ibid.

50 T. J. Pempel (1982) Policy and Politics in Japan, Philadelphia: Temple University Press, pp. 13–20.

Chapter 2 Definition of biotechnology

1 *A Supplement to the Oxford English Dictionary*, vol. 1, Oxford University Press, Oxford, 1972, p. 267: *Science*, 105: 217.

2 'Ergonomics' itself first appeared in 1949, and has probably gained its recent popularity because of its similarity to the word 'economics'.

3 *A Supplement to the Oxford English Dictionary*, p. 267.

4 Margaret Sharp (1984) *The New Biotechnology: European Governments in Search of a Strategy*, Brighton: Sussex University, p. 14. Peter Daly (1985) *The Biotechnology Business – The Strategic Analysis*, London: Frances Pinter, p. 6. The word 'generations' first appeared in the Spinks Report (ACARD 1980) and is generally attributed to Dr Spinks himself.

5 I. J. Higgins, D. J. Best, and J. Jones (eds) (1985) *Biotechnology – Principles and Applications*, Oxford: Blackwell Scientific Publications, p. 14; and Sharp (1984), p. 14.

6 Higgins (1985), p. 7.

7 ibid. p. 6.

8 *Tokyo Daigaku Nōgakubu Nōgeikagakuka Gaidansu Bukku* 1986

Notes

(Guide Book to Tokyo University Agricultural Faculty, Department of Agricultural Chemistry), Tokyo: Tokyo University, p. 1.

9 Alun Anderson (1984) *Science and Technology in Japan*, London: Longman, p. 110.

10 Kamibayashi Akira (1982) *Sangyō no Nyūfuronteia Baiotekunorojī (Industry's new Frontier – Biotechnology)*, Tokyo: MITI, p. 18.

11 Sharp (1984) p. 49; also see Gene Gregory (1985) *Japanese Electronics Technology: Enterprise and Innovation*, Tokyo, Japan Times.

12 *Nature*, 1953, 171, 737.

13 *Commercial Biotechnology: An International Analysis* (1984) Washington, DC, US Congress, Office of Technology Assessment, p. 3.

14 *21 Seiki o Hiraku Baioindasutorii – Sono Tenbō to Kadai – (Bioindustry Opening Up the 21st Century – Prospects and Problems)*, Tokyo: MITI's Basic Industry Section, 1984, p. 3.

15 *1970 Nendai ni Okeru Sōgōteki Kagaku Gijitsu Seisaku no Kihon ni Tsuite (The Foundation of a Comprehensive Science Policy for the 1970s)*, Tokyo: Council of Science and Technology, 1971, p. 37.

16 *Seio 2000 Nen ni Okeru Baiotekunorojī no Sangyō Kōzō ni Obosu Inpakuto (Impact of Biotechnology on Industrial Structure in the Year 2000)*, Tokyo: Japanese Fermentation Association, Bioindustry Promotion Department, 1985, August, p. 6.

17 The term NBFs was first used in the OTA report of January 1984, p. 6, and was defined as 'entrepreneurial ventures started specifically to pursue applications of biotechnology'.

18 Saitō Hyūga (1985) Biotechnology R&D: Japan and the World, *Science: Technology in Japan*, April/June, p. 8.

19 Imada, (1986) Baiotekunorojī Sangyō no Gendai to Tenbō (The Present and Future Prospects of the Biotechnology Industry), *Sangyō Richi (Journal of Industrial Location)*, 25(5): 9–11.

20 *Baioindasutorī Shinkō Iinkai Hōkokusho (The Bioindustry Advisory Committee (BAC) Report)*, Tokyo: MITI, Aug. 1984, p. 2.

21 Higgins (1985) p. 24.

22 BAC Report, p. 5.

23 Steve Prentis (1984) *Biotechnology: A New Industrial Revolution*, London: Orbis Publishing, p. 177.

24 ibid. p. 85.

25 ibid. p. 115.

26 Kamibayashi Akira (1982) p. 80.

27 John Hartley (1984) Japan's rising star. *Engineer*, 258, 36.

28 Prentis (1984) p. 86.

29 *Nikkei Baioteku (Nikkei Biotech)*, 13 January 1986.

30 Prentis (1984) p. 88.

31 ibid. p. 74.

32 ibid. p. 73.

33 6–APA is produced simply by hydrolysis of penicillin which itself is produced by fermentation.

34 Higgins (1985) pp. 9–11.
35 BAC Report, p. 6.
36 ibid.
37 Prentis (1984) p. 128.
38 Higgins (1985) p. 97.
39 Prentis (1984) p. 131.
40 *New Scientist*, Dec. 1985.
41 Prentis (1984) p. 87.
42 ibid. p. 25.
43 ibid. p. 179.
44 ibid. p. 147.
45 Bernard Dixon (1985) Biotechnology: genes out of control, *New Scientist*, 24 October.
46 Uchida Hisao, Speech to Tokyo International Biofair 1986, *Kogyo Ryo ni Kanshite-Nihon no Jirei Kenkyu*, Japan's Actual Research – in Industrial Application (of Biotechnology), Tokyo: 16 October 1986.

Chapter 3 Building a high tech society in Japan

1 Nikkei Science Company (1985) *Nihon no Sentan Gijitsu – 21 Seiki e no Tenbō* – (Japan's High Technology – Prospects for the 21st century), Tokyo: Mita Press, p. 1.
2 Research and Development (R&D) is usually reduced analytically to three components: basic research, applied research, and development research. Basic research is research aimed at understanding the material under study rather than the practical application thereof. Applied research is research directed towards practical application of knowledge. Development research is research aimed at the systematic use of knowledge for production of useful materials, devices, systems, methods, processes, design, and the development of prototypes.
3 Takano Hajime (1980) *Tsusanshō no Yabō* (MITI's Ambition), Tokyo: Nikkan Kogyo Newspaper Press, p. 26.
4 Yano Toshiko and Amadani Naohiro (1984) *Haitekunorojī No Sōzō to Katsyō* (The Creativity and Practical Use of High Technology), Tokyo: Daiichi Hoki, p. 2.
5 See Yano and Amadani; Murakami Yasusuke (1986) Technology in transition: two perspectives on industrial policy, in *Japan's High Technology Industries*, Seattle: University of Washington Press, pp. 211–43.
6 Ishi Makoto and Nagaoka Aki (1985) *21st Seiki e no Dōhyō* (The Road to the 21st Century), Tokyo: Japan Science Foundation, p. 30.
7 ibid. p. 31.
8 Yano and Amadani (1984) p. 15.
9 Ōtsuki Shōten (1985) *Tekunoporisu to Chiiki Kaihatsu* (Technopolis and Regional Development), Tokyo: Nihon Kagakusha Kaigi Press, p. 37.
10 ibid.

Notes

11 See Graham Allison (1971) *Essence of Decision: Explaining the Cuban Missile Crisis*, Boston: Little, Brown, p. 37.
12 Ōtsuki Shōten (1985) p. 34.
13 *Shimon Dai ll Gō*: Council on Science and Technology, p. 8.
14 *Nature*, 317, 24 Oct. 1985.
15 *Nikkei Shimbun Yūkan* (Japan Daily Financial Evening News), 28 March 1983.
16 Motojima Naoki interview, 21 Oct. 1986.
17 Michael Rogers (1982) The Japanese Government's role in biotechnology R&D, *Chemistry and Industry*, 7 Aug., p. 534.
18 *Shimon Dai ll Gō*: Council on Science and Technology, p. 101.
19 ibid. pp. 101–9.
20 ibid. p. 101.
21 ibid. pp. 107–8.
22 Teretopia Kōsō no Suishin ni Tsuite (Concerning the Promotion of the Teletopia Concept), internal document of the Ministry of Posts and Telecommunications, unpublished.
23 Refers to the industrial belt stretching from Tokyo to Osaka along the Pacific south coast of Hokkaido.
24 Ōtsuki Shōten (1985) p. 34.
25 *Shimon Dai ll Gō*: p. 9.
26 Kurita interview, Oct. 1985.
27 Yano and Amadani (1984) p. 31.
28 Gary Saxonhouse (1986) Industrial policy and factor markets: biotechnology in Japan and the US *Pacific Economic Papers*, no. 136, pp. 50–3.
29 Science and Technology Agency (1984) *Summary of White Paper 1983*, Tokyo: Science and Technology Agency, p. 14.
30 Ronald Dore (1983) *A Case Study of Technology Forecasting in Japan, the Next Generation Base Technologies Development Programme*, London: The Technical Change Centre, p. 5.
31 ibid. pp. 7–9.
32 Jon Elster (1983) *Explaining Technical Change*, Cambridge: Cambridge University Press, p. 107.
33 ibid. p. 120.
34 E. Mansfield (1968) *The Economics of Technical Change*, New York: Norton, p. 10. A very useful definition also adopted by P. S. Johnson (1975) *The Economics of Invention and Innovation*, London: Martin Robertson and Peter Daly (1985) *The Biotechnology Business – A Strategic Analysis*, London: Frances Pinter, p. 54.
35 J. Schmookler (1966) *Invention and Economic Growth*, Cambridge, Mass.: Harvard University Press, p. 3.
36 P. S. Johnson (1975) *The Economics of Invention and Innovation*, London: Martin Robertson, p. 18.
37 Nathen Rosenberg (1983) *Inside the Black Box: Technology and Economics*, Stanford: Stanford University Press, p. 143.
38 A. R. Hall (1963) *From Galileo to Newton, 1630–1720*, London: Collins, quoted in Rosenberg (1983) p. 13.

39 Rosenberg (1983) p. 156.
40 ibid. p. 159.
41 Moses Abramovitz (1956) *American Economic Review Papers and Proceedings*; and Robert Solow (1957) *A Review of Economics and Statistics*, quoted in Rosenberg (1983) pp. 10–15.
42 Rosenberg (1983) pp. 23–24.
43 ibid. p. 57.
44 ibid. pp. 60–1.
45 Saitō Hyūga interview, 14 Aug. 1986.
46 Moritani Masanoru (1982) *Japanese Technology*, Tokyo: Simul Press, p. 114.
47 Rosenberg (1983) p. 187.

Chapter 4 Strategies for developing biotechnology in Japan

1 Two minor central government offices, the Construction Ministry and the Environment Agency are also involved with biotechnology. Since their contribution to overall biotechnology policy in Japan has been minimal, they are not discussed in this book.
2 Oita Takahisa interview.
3 *Shimon Dai Gogō 1970 Nendai ni Okeru Sōgoteki Kagakugijutsu Seisaku no Kihon ni Tsuite ni Tai suru Tōshin. (The Fifth Recommendation Concerning the Basis of a Comprehensive Science and Technology Policy in the 1970s)*. Tokyo: Council on Science and Technology, 1971, p. 1.
4 *Kagaku Gijutsu Kaigi no Gaiyō (An Outline of the CST)*, Tokyo: Council on Science and Technology, Dec. 1985, p. 1.
5 *Nikkei Shimbun (Japanese Financial Times)*, 1 March 1982.
6 ibid.
7 *Kagaku Gijutsu Gaiyō* pp., 32. Two other *Bukai* committees were also formed that day, the Energy Science and Technology and the Research Targeting *Bukai* Committees.
8 ibid. p. 5.
9 *Keidanren 1986* (Pamphlet), July 1986, Public Affairs Dept, *Keidanren*.
10 Originally both the MOE and the STA were in charge of the CST, a situation that has changed drastically over the years.
11 *Laifusaiensu* STA published Pamphlet 1985.
12 *Baio Gyōsei o Ōu* (1986) Tokyo: Nihon Kogyo Shinbun, p. 27.
13 *Commercial Biotechnology: An International Analysis* (1984) Washington, DC: US Congress Office of Technology Assessment, p. 476.
14 Tahara Sōichirō (1981) *Idenshi Sangyō Kakumei (Revolution of Industrial Genetics)*, Tokyo: Bungei Shinju, pp. 142–3.
15 ibid. pp. 133–4.
16 Akira Matsuda (1982) Wagakuni no Baiotekunoroji Seisaku (Biotechnology Policy in Japan) *Kagaku to Kogyo*, 35 (11): 773.
17 Alan T. Bull, Geoffrey Holt and Malcolm D. Lilly (1982)

Notes

Biotechnology: International Trends and Perspectives, OECD, pp. 62–3.
18 Morishita Noboru interview 20 Oct. 1986.
19 Morishita Noboru interview; Takano Hajime (1980) *Tsusan no Yabo (MITI's Ambition)*, Tokyo: Nikkan Kogyo Newspaper Press, p. 146.
20 Takano (1980) p. 143.
21 In the first half of January alone, numerous headlines appeared such as, 'Stocks Related to Genetic Engineering Lead the New Spring Market' (*Nikkan Investment Newspaper*, 8 January 1981), 'Genetic Engineering Suddenly Booms' (Special supplement to *The Stocks Newspaper*, 8 January), 'The Life Science of True Flotation – Toray Mitsubishi Chemical, and Takara' (*The Stocks Newspaper*, 13 January), 'Biotechnology Conceives a Huge Dream – Potential for Japanese Mining, Biomass and Interferon' (*Nikkan Investment Newspaper*, 16 January) and so on.
22 *Nature*, 314, 395, April 1985; *Nature*, 307, 504, Feb. 1984.
23 *Nikkan Kōgyō Shimbun*, 14 Dec. 1982.
24 Saitō Hyùga (ed.) (1984) *Mirai Sangyō o Umu Baiotekunorojī (Biotechnology Giving Birth to Future Industries)*, Tokyo: MITI, pp. 195–6.
25 *Research and Development Project of Basic Technology for Future Industries*, Agency of Industrial Science and Technology, Ministry of International Trade and Industry, 1986, p. 1.
26 Takano (1980) pp. 32–3.
27 ibid. p. 33.
28 ibid. p. 28; Arai Takashi (1984) Jisedai Sangyō Kikan Gijutsu Kenkyu Kaihatsu no Suishin (Promotion of R&D in the Next Generation Basic Technology Project), *Tsūsanshō Kōhō (MITI's Report)*, 27 Nov. p. 11.
29 Takano (1980) p. 26.
30 Ikegaya Soichi interview, 17 Sept. 1986.
31 Ronald Dore (1983) *A Case Study of Technology Forecasting in Japan, the Next Generation Base Technologies Development Programme*, London: Technical Change Centre, p. 12.
32 ibid. pp. 11–12.
33 ibid.
34 ibid. p. 8.
35 ibid.
36 Sasaki Yoshihiko interview, 26 Sept. 1986.
37 Dore (1983) p. 18.
38 Alfred Spinks (1980) *Biotechnology: Report of a Joint Working Party*, London: Advisory Board for the Research Councils and the Royal Society.
39 Government White Paper: *Biotechnology*, Department of Industry, 1981.
40 Dore (1983) p. 15.
41 Sasaki interview.
42 *New Scientist*, 21 March 1985, p. 32.

43 Gary Saxonhouse (1986) Industrial policy and factor markets: biotechnology in Japan and the US *Pacific Economic Papers*, no. 136, Canberra: Australia–Japan Research Centre, p. 24.
44 *Nihon Keizai Shimbun*, 12 Aug. 1981.
45 Tahara (1981) p. 149.
46 ibid.
47 *Japan Bioindustry Letters – Special Issue*, Tokyo: BIDEC, 1985.
48 'Magic Cane' (*mahō no tsue*) and 'Even the Cat and the Ladle' (*neko mo shakushi mo*).
49 Saxonhouse (1986) p. 6.
50 *Nihon Keizai Shimbun*, 28 May 1984.
51 *Nihon Keizai Shimbun*, 22 June 1984.
52 Ikegaya Soichi interview, 17 Sept. 1986.
53 Ikegaya Soichi interview, 30 Oct. 1986.
54 Kaigi (1985) p. 105.
55 Robert Fujimura (1986) Unclassified Telex from Tokyo to Washington DC.
56 Saitō Hyūga interview 3 Oct. 1986.
57 Uchida Hisao (1986) Speech to BioFair '86, Tokyo: 16 Oct. *Kōgyōryō ni Kanshite – Nihon no Jirei Kenkyū* (Japan's Actual Research – Concerning Industrialisation).
58 *Commercial Biotechnology* OTA Report, p. 555.
59 Miyata Mitsuru interview.
60 Uchida Speech.
61 He argues that molecular genetic studies of eukaryotes were well advanced, but rDNA cloning of eukaryote genes was not. Professor Saitō Hyūga, also of Tokyo University, seems to be more skeptical contending that university researchers had long been interested in rDNA technology.
62 Uchida Speech.
63 ibid.
64 ibid.
65 ibid.
66 Saitō Hyūga (1980) Kumikae DNA Jikken Shishin no Sono Ato (After the Recombinant DNA Guidelines), *Kagaku to Seibutsu* (Chemistry and Biology), June.
67 *Nikkei Shimbun*, 28 May 1984.
68 Frank Young (1986) Speech to BioFair '86, Tokyo: 15 Oct. *America's Bioindustries*.
69 Uchida Speech.
70 ibid.

Chapter 5 Implementation of biotechnology policy – strategy for its promotion

1 Gary Saxonhouse (1986) Industrial policy and factor markets: Biotechnology in Japan and the United States, *Pacific Economic Papers*, no. 136, June, p. 7.

Notes

2 ibid. p. 10.
3 ibid. p. 11.
4 Miyata Mitsuru interview, 31 Oct. 1986.
5 *Commercial Biotechnology; An International Analysis*, Washington DC: US Congress, Office of Technology Assessment, 1984, pp. 505–30.
6 Kurata Kenji interview, Sept. 1985.
7 Yoshida Motoki (1984) *Reputo: Baiotekunorojī no Shinkaihatsu to Kadai* (*Report: The Question of Biotechnology's New Development*), Tokyo: Japan Industrial Bank, p. 32.
8 'Responsive dependence' used by Gary D. Allinson (1975) *Japanese Urbanism: Industry and Politics in Kariya, 1872–1972*, Berkeley; University of California Press, pp. 34–35, quoted in Chalmers Johnson (1981) *MITI and the Japanese Miracle*, Stanford: University of Stanford Press, p. 24.
9 Daniel Okimoto (1986) Regime characteristics of Japanese industrial policy, in Hugh Patrick (ed.) *Japan's High Technology Industries*, Seattle: University of Washington Press, p. 70.
10 Saxonhouse (1986) pp. 26–7.
11 Kosai Yutaka and Ōgino Yoshitarō (1984) *Contemporary Japanese Economy*, Tokyo: Macmillan Press, p. 95.
12 Saxonhouse (1986) p. 36.
13 Clem Tisdell (1975) An Australian review of Japanese science and energy policy, in *Australia–Japan Economic Relations Project*, Canberra: National University Canberra, p. 16.
14 *San. Kan. Gaku. no Kenkyū Kyōryoku ni Kakaru Shisaku no Genjō* (Present Situation of Policies Concerning Co-operative Research Between Industry, Government and Universities), Internal MOE Document July 1986.
15 *Baio Gyōsei o Ōu (Following Biotechnology Administration)*, Tokyo: Nihon Kigyo Shimbun, 1986, p. 21.
16 Saxonhouse (1986) p. 36.
17 Alun Anderson (1984) *Science and Technology in Japan*, London: Longman, p. 57.
18 Tanaka Masami interview, 12 Sept. 1986.
19 Tahara Sōichirō (1981) *Idenshi Sangyō Kakumei* (*Industrial Revolution in Genetics*), Tokyo: Bungei Shinj, p. 147.
20 Margaret Sharp (1984) *The New Biotechnology: European Governments in Search of a Strategy*, Brighton: University of Sussex, p. 64.
21 ibid. p. 69.
22 Miyata Mitsuru interview.
23 *Nikkei Sangyo Shimbun*, 4 Sept. 1982.
24 Miyata Mitsuru interview.
25 ibid.
26 Okimoto (1986) p. 65.
27 Shimizu Makoto interview, 16 Sept. 1986.
28 Alun Anderson (1984) p. 235.

Notes

29 Tanaka Masami interview.
30 P. S. Johnson (1975) *The Economics of Invention and Innovation*, London: Martin Robertson, pp. 190–5.
31 Kent Calder Lecture, Princeton University, Spring 1984.
32 Sasaki Shuichi interview, 26 Sept. 1986.
33 Tanaka Masami interview.
34 *Baioindasutorī Seisaku no Suishin (Promotion of Bioindustry Policy)*, Tokyo: MITI's Basic Industry Section, 1985, p. 13.
35 Nikkei Baioteku, January Issues from 1982 to 1986.
36 Yoshida Motoki (1984) Reputo: Baiotekunoroji no Shinkaihatsu to Kadai (The Question of Biotechnology's New Development), Tokyo: Japan Industrial Bank, p. 22.
37 Sasaki interview, *Nikkei Baioteku*, 12 Oct. 1985, p. 5.
38 Edward A. Feigenbaum and Pamela McCorduck (1983) *The Fifth Generation: Artificial Intelligence and Japan's Computer Challenge to the World*, Reading, Mass.: Addison-Wesley, p. 109.
39 *Kagaku Kōgyō Nippo*, 27 April 1983.
40 Tanaka Masami, *Regional Development and Biotechnology*, MITI internal document.
41 Nakata Naoki interview, 22 Sept. 1986.
42 Morishita Noboru interview, 28 Aug. 1986.
43 *Baio Gyōsai o Ōu*, p. 14.
44 Aiba Yasuhide interview, 24 Sept. 1986.
45 Tanaka Masami interview, 12 Sept. 1986.
46 *Keidanren Review*, no. 95, Oct. 1985, p. 9.
47 *Baio Gyōsei o Ōu*, p. 7.
48 Saxonhouse (1986) p. 53.
49 Saxonhouse (1986) p. 49; Murakami Yoichirō interview.
50 Ōyama Chō interview, 11 Sept. 1986.
51 *Biotechnology in Japan*, SERC Directorate, May 1987, p. 2.
52 Fujimura Robert interview, 21 Oct. 1986.
53 *Nature*, 323, 284, 1986.
54 ibid.
55 Kobayashi Shinichi interview, 11 Sept. 1986.
56 Sharp (1984) pp. 90–1. Also see US Government 919830 *Interagency Working Group on Competitive Transfer Aspect of Biotechnology Report*.
57 Fujimura Robert interview, 1 Sept. 1986.
58 John Beggs and Pamela Fayle (1985) A comparative case study of the biotechnology industries in Australia and Japan, *The Pacific Economic Papers*, no. 125, Aug.
59 Saitō Hyūga interview, 3 Oct. 1986.
60 Saitō Hyūga (1985) Biotechnology R&D; Japan and the World, *Science and Technology*, April/June, pp. 10–11.
61 Saitō Hyūga interview, 14 Aug. 1986.
62 Tahara Soichi (1981) Idenshi Sangyo Kakumei (Revolution of Industrial Genetics), Tokyo; Bungei Shinju, p. 158.
63 *Nikkei Baioteku (Nikkei Biotech)* 12 April 1982.

143

Notes

64 *Nikkei Baioteku* (*Nikkei Biotech*) 26 Oct. 1986.
65 Saitō Hyūga (1982) Idenshi Sōsaku ni Okeru Seibutsuteki Fūjikome
no Anzensei ni Tsuite (Concerning the Safety of Biological
Containment in Genetic Manipulations) *Hakkō to Kōgyō*
(*Fermentation and Industry*), 40 (3), 30.
66 *Nikkei Baioteku* (*Nikkei Biotech*) 26 April 1982.

Conclusion: The Triple N Synthesis

1 Daniel Okimoto (1986) Regime characteristics of Japanese industrial
policy, in Hugh Patrick (ed.) *Japan's High Technology Industries*,
Seattle: Washington University Press, pp. 35–97.
2 All references to *Nebukai* hereafter pertain to the new differentiated
model featuring the academics.
3 Satō Masuke interview, 9 Sept. 1986.
4 Okimoto (1986) p. 39.
5 James E. Anderson (1975) *Public Policy-Making*, New York:
Praeger, p. 2.
6 *21 Seiki o Hiraku Baioindasutorī – Sono Tenpō to Kadai –* (*Opening
up the 21st Century with Bioindustry – Prospects and Problems –*),
Tokyo: MITI's Basic Industry Section Press, 1984, p. 37.
7 Saitō Hyūga interview, 14 Aug. 1986.
8 Graham Allison (1971) *Essence of Decision: Explaining the Cuban
Missile Crisis*, Boston: Little, Brown, p. 157.

Bibliography

Abegglen, James (1970) 'The Economic Growth of Japan', *Scientific American*, 222.

Akira Matsuda (1982) 'Wagakuni no Baiotekunorojī Seisaku' (Biotechnology Policy in Japan) (*Kagaku to Kōgyō*), 35, no. 11.

Allen, G. C. (1981) *The Japanese Economy*. London: Weidenfeld & Nicolson.

Allison, Graham T. (1971) *Essence of Decision: Explaining the Cuban Missile Crisis*, Boston: Little, Brown & Company.

Anderson, Alun (1984) *Science and Technology in Japan*. London: Longman.

Anderson, James E. (1975) *Public Policy-Making*. New York: Praeger Publishers.

Arai Takashi (1985) 'Jisedai Sangyō Kikan Gijitsu Kenkyū Kaihatsu no Suishin' (Promotion of R&D in the Next Generation Basic Technology Project), in *Tsūsanshō Kōhō* (*MITI's Report*) 27 November.

Asahi Nenkan 1980–1985 (*Asahi Yearbooks 1980–1985*). Tokyo: Asahi Publishing Company.

Baio Gyōsei o Ōu (*Following 'Bio' Policy*). Tokyo: Nihon Kogyo Shimbun, 1986.

Bazelon, David L. (1986) 'Governing technology: values, choices and scientific progress' in Joseph G. Perpich (ed.) *Biotechnology in Society*. New York: Pergamon Press.

Beetham, David (1987) *Bureaucracy*. Milton Keynes: Open University Press.

Beggs, John and Fayle, Pamela (1985) 'A Comparative case study of the biotechnology industries in Australia and Japan' in *Pacific Economic Papers* no. 125.

Biotechnology Department of Industry (White Paper) British Government.

Biotechnology. The Last Word, Tanaka Masami in *Biotechnology* February 1985.

Blumenthall, Tuvia (1973/4) 'Japan's technological strategy' in *Australia, Japan Economic Relations Research Project*. Canberra: Australian National University.

Bibliography

Bull, Alan, Holt, Geoffrey and Lilly, Malcolm (1982) *Biotechnology: International Trends and Perspectives*.

Cape, Ronald E. (1986) 'Future prospects in biotechnology: A challenge to United States leadership' in Joseph Perpich (ed.) *Biotechnology in Society, Private Initiatives and Public Oversight*. New York: Pergamon Press.

Chemistry and Industry.

Commercial Biotechnology: An International Analysis. Washington, DC: US Congress, Office of Technology Assessment, January 1984.

Council of Science and Technology (1985) *Kagaku Gijitsu Kaigi no Gaiyō* (An Outline of the CST) Tokyo: CST, December.

Crozier, Michael (1964) *The Bureaucratic Phenomenon*. London: Tavistock Publications.

Daley, Peter (1985) *The Biotechnology Business*. London: Frances Pinter (Publishers) Ltd.

Dore, Ronald (1983) *A Case Study of Technology Forecasting in Japan: The Next Generation Base Technologies Development Programme*. London: Technical Change Centre.

Dougherty, James E. and Pfaltzgraff, Robert L. (1981) *Contending Theories of International Relations*. New York: Harper & Row.

Elster, Jon (1983) *Explaining Technical Change*. Cambridge: Cambridge University Press.

Feigenbaum, Edward A. and McCorduck, Pamela (1983) *The Fifth Generation: Artificial Intelligence and Japan's Computer Challenge to the World*. Reading, Mass.: Addison-Wesley.

Fukui, Saburo *et al.* (1986) *Baio no Sekai* (*The World of 'Bio'*). Tokyo: Nikkan Kōgyō Shimbun Company.

Fukuoka Planning and Development Section of MITI (1986) *Baioindasutorii in Kyushu* (*Bioindustry in Kyushu*). Fukuoka: Kyushu's Industrial Technology Centre.

Genetic Engineering and Biotech Monitor. Vienna: United Nations Industrial Development Organization, 1980–7.

Gijitsu Dōyūkai. Gijitsu Kakushin No Sokushin Ni Kan Suru.

Gregory, Gene (1985) *Japanese Electronics Technology: Enterprise and Innovation*. Tokyo: The Japan Times. 'Japan's Bioindustry' in *New Scientist* 1982.

Gyōsei Kikozu (*Administrative Charts*). Tokyo: Administration and Management Research Centre, 1980.

Harari, Ehud, *The Institutionalization of Policy Consultation* (unpublished).

Hartley, John (1984) 'The Borrowers', *Engineer*, 17 May.

Higgins, I. J., Best, D. J. and Jones, J. (1985) *Biotechnology – Principles and Applications*. Oxford: Blackwell Scientific Publications.

Horizons in Biotechnology.

Isao Karube, Shotarō Kohtsuki, Isao Endo (1985) *Baio no Chosen* (*The Bio Challenge*). Tokyo: Kodansha.

Ishi, Makoto and Nagaoka, Aki (1985) *21 Seiki e no Dōhyō* (*The Road to the 21st Century*). Tokyo: Japan Science Foundation.

Jacobsson, S., Jamison, A. and Rothman, H. (eds.) (1986) *The Biotechnological Challenge*. Cambridge: Cambridge University Press.
Japan Bioindustry Letters – Special Issue. Tokyo: BIDEC, 1985.
Japan Industrial Bank (1985) *Nihon Sangyō no Dokuhon (The Companion to Japan's Industry)*, Tokyo: Tōyō Keizai Shimpō.
JEI Report: *Update on Japan's Biotechnology Industry: Stunted Growth?* Japan Economic Institute No. 5A, February 1986.
Johnson, Chalmers (1982) *MITI and the Japanese Miracle*. Stanford: Stanford University Press.
Johnson, P. S. (1975) *The Economics of Invention and Innovation*. London: Martin Robertson.
Kagaku kōgyō Nippō (Chemical Industry Newspaper)
Kamibayashi, Akira (1982) *Sangyō no Niyūfuronteia Baiotekunorojī (Industry's New Frontier Biotechnology)*. Tokyo: MITI.
Keidanren 1986 July 1986, Public Affairs Dept., Keidanren.
Kōdōgijitsu Ni Rikkyakushita Kōgyō Kaihatsu Ni Kan Suru Keikaku (A Plan Concerning Industrial Development Based on Advanced Technologies), Tokyo: Okayama Prefecture, 1983.
Kosai, Yutaka and Ōgino, Yoshitarō (1984) *Contemporary Japanese Economy*. London: Macmillan.
Lewis, Herman (1985) 'Biotechnology in Japan', in *Scientific Bulletin*, 10 (2), Dept. of the Navy Office of Naval Research Far East April–June.
Mansfield, E. (1968) *The Economics of Technical Change*. New York: Norton.
March, James and Simon, Herbert (1958) *Organisations*. New York: John Wiley.
Ministry of Agriculture, Forestries, and Fisheries (1986) *Nōrinsuisan Gijitsu Kaigi 30 Nen (Agriculture Technology Council 30 Years)*. Tokyo: MAFF.
Ministry of Education, *San. Kan. Gaku no Kenkyū Kyōkyoku ni Kakaru Shisaku no Genjo (Present Situation of Policies Concerning Cooperative Research Between Industry, Government and Universities)* July 1986 Internal MOE Document.
MITI (1985/6) *Baioindasutorī Seisaku No Suishin (Promotion of Bioindustrial Policies)*. Tokyo: MITI's Basic Industries Bureau.
MITI (1984) *Baioindasutorī Shinko Iinkai Hokokusho (Report of the Bioindustry Advisory Council)*. Tokyo: MITI, August.
MITI (1984) *Jisedai Sangyō kiban Gijitsu Kenkyū Kaihatsu Seido (Next Generation Basic Technologies R&D System)*. Tokyo: AIST, October.
MITI (1984) *21 Seki o Hiraku Baioindasutorī (Bioindustry Opening Up the 21st Century)*. Tokyo: Basic Industries Bureau.
MITI (1986) *21 Seki Ni Muketa Gijitsu Kaihatsu to Kokusai Kōryū No Arikata (Technological Development and International Transfer For the 21st Century)* Interim Report. Tokyo: AIST.
Miyakawa, Hiroshi (1985) 'Present and future of new media in Japan' in *Science and Technology in Japan*. Tokyo: Japan Times.
Morishita, Noboru (1985) 'Shingikai Ni "Shingi Ranpu" Tento' (Lighting a 'Deliberative Lamp' on Deliberation Councils) in *Tsūsan Jyanaru* November.

Bibliography

Moritani, Masanori (1982) *Japanese Technology – Challenge of Creativity*. Tokyo: Simul Press.

Nature

New Scientist

Nihon Kingendaishi Jiten. Tokyo: Toyo Keizai Shinposha, 1978.

Nihon Kōgyō Shimbun (Japan Industrial Newspaper).

Nihon Kōgyō Shimbun Corps. *Baio Gyōsei o Ōu (Following Biotechnology Administration)*. Tokyo: Nihon Kōgyō Newspaper Press, 1986.

Nihon Kingendaishi Jiten (A Dictionary of Recent, Modern Japanese History). Tokyo: Toyo Keizai Shinpo Limited, 1978.

Nihon Kagakusha Gaiki (1985) *Tekunoporis to Chiiki Kaihatsu (Technopolis and Regional Development)*. Tokyo: Otsuki.

Nihon No Sentan Gijitsu (Japan's High Technology). ed. Nikkei Science Tokyo: Mita Publications, 1985.

Nikkan Kōgyō Shimbun (Japanese Industrial Daily).

Nikkan Special Bio Corps (1986) *Baio no Sekai (The World of Bio)*. Tokyo: Nikkan Kogyo Shimbun.

Nikkei Baiotekku. (Nikkei Biotech).

Nikkei, Baiotekunorojī Saizensen (Biotechnology, The Front Line). Tokyo: Nikkei Shimbun, 1984.

Nikkei Baiotekunoroji Saishin Yōgo Jiten (Nikkei's Biotechnology New Word Dictionary). Tokyo: Nikkei Biotech, 1985.

Nikkei Keizai Shinbun (Financial Times of Japan). Tokyo.

Ochi, Yuichi (1972) 'The organization and activities of the Science Council of Japan' *Nature*, 240.

Okabayashi, Tetsuo (1986) 'Gyōsei Kara Mita Baiotekunorojī no Genjo to Kogō' (The present and future of biotechnology: from a policy perspective) in *BioFair Tokyo '86 – Guidebook*. Tokyo: BIDEC.

Office of Technology Assessment (1984) *Commercial Biotechnology: An International Perspective*. Washington: US Congress.

Otsuki Shoten (1985) *Tekunoporisu to Chiiki Kaihatsu (Technopolis and Regional Development)*. Tokyo: Nihon Kagakusha Kaigi Press.

Pacific Economic Papers.

Panem, Sandra (ed.) (1985) *Biotechnology and Public Policy*. Washington, DC: Brookings.

Patrick, Hugh (ed.) (1986) *Japan's High Technology Industries*. Seattle: University of Washington Press.

Pempel, T. J. (1982) *Policy and Politics in Japan*. Philadelphia: Temple University Press.

Pempel, T. J. (ed.) (1977) *Policymaking in Contemporary Japan*. Ithaca: Cornell University Press.

Perpich, Joseph G. (1986) *Biotechnology in Society – Private Initiatives and Public Oversight*. New York: Pergamon.

Prentis, Steve (1984) *Biotechnology: A New Industrial Revolution*. London: Orbis.

Raifusaiensu No Suishin Ni Kan Suru Kenkai – Jitsuyōka No Sokushin Ni Mukete – (Understanding the Progress of the Life Sciences – Towards Actual Progress –). Tokyo: *Keidanren* Report, 1985.

Research and Development Project of Basic Technology for Future Industries, Agency of Industrial Science and Technology, Ministry of International Trade and Industry 1986.

Rogers, Michael (1982) 'The Japanese Government's role in biotechnology R&D', *Chemistry and Industry*, 7 August.

Rosenberg, Nathen (1983) *Inside the Black Box: Technology and Economics*. Stanford: Stanford University Press.

Ruttan, Vernon (1959) 'Usher and Schumpeter on invention, innovation and technical change', *Quarterly Journal of Economics*.

Saitō Hyūga (1980) 'Kumikae DNA Jikken Shishin no sono ato' ('After the recombinant DNA Guidelines') *Kagaku to Seibutsu*, June.

Saitō Hyūga (1982) 'Idenshi Sōsaku ni Okeru Seibutsuteki Fujikome no Anzensei ni Tsuite' (Concerning the Safety of Biological Containment in Genetic Manipulations), *Hakkō to Kōgyō (Fermentation and Industry)*, 40 (3).

Saitō, Hyūga (ed.) (1984) *Mirai Sangyo o Umu Baiotekunorojī (Biotechnology Giving Birth to Future Industries)*. Tokyo: MITI.

Saitō, Hyūga (1985) 'Biotechnology R&D: Japan and the World' Science & Technology in Japan.

Saxonhouse, Gary (1986) 'Industrial policy and factor markets: biotechnology in Japan and the US', in *Pacific Economic Papers* No. 136.

Schmid, R. D. (1984) 'Biotechnology in Japan – A mini review', *Applied Microbiology and Biotechnology*, 22.

Schmookler, J. (1966) *Invention and Economic Growth*. Cambridge, Mass.: Harvard University Press.

Science and Technology Agency (1984) *White Paper on Science and Technology 1983: Towards Creation of New Technology for the 21st Century (Summary)*. Tokyo: Science and Technology Agency.

Science and Technology Agency (1985) *Raifusaiensu* (Pamphlet). Tokyo: STA.

Science and Technology Agency *Shimon Dai Gogo 1970 Nendai no okeru Sōgōteki Kagakugijitsu Seisaku no Kihon ni tsuite ni tai suru toshin*. The Fifth Recommendation Concerning the Basis of a Comprehensive Science and Technology Policy in the 1970s.

Science and Technology Agency (1984) *Shimon Dai 11 Go. 'Aratana Josei Ilenka Ni Taio shi, Chokiteki Tenbō Ni Tatta Kagaku Gijitsu Shinkyo No Sōgōteki Kihon Hōsaku Ni Tsuite, Ni Taisuru Kaishin'. (The 11th Recommendation of the Council on Science and Technology entitled, 'A Comprehensive and Fundamental Policy For the Promotion of Science and Technology to Focus on Currently Changing Situations From a Long-term Perspective'.)* Tokyo: Science and Technology Agency.

Science and Technology Agency (supervised) (1984) 21 Seki e no Kōzō (Creativity for the 21st Century). Tokyo: Jihyosha.

Science Council of MOE (1986) *Daigaku Nado Ni Okeru Baiosaiensu Kenkyū No Suishin Ni Tsuite (Kengi). (The Promotion of Bioscience Research in Universities and other Academic Institutions)*. Tokyo: MOE.

149

Bibliography

Sharp, Margaret (1984) *The New Biotechnology: European Governments in Search of A Strategy.* Sussex: University of Sussex.

Spinks Report (1980) *Biotechnology: Report of a Joint Working Party.* ACARD: London.

Stockwin, J. A. A. (1982) *Japan: Divided Politics in a Growth Economy*, 2 edn. London: W. W. Norton & Company, p. 154.

Tahara, Soichirō (1981) *Idenshi Sangyō Kakumei (Genetic Industrial Revolution).* Tokyo: Bungei Shunju.

Takano, Hajime (1980) *Tsūsanshō no Yabo (MITI's Ambition).* Tokyo: Nikkan Kogyo Shinbun.

Tanaka, Masami, *Regional Development and Biotechnology* (MITI internal document).

Teretopia Kōsō no Suishin ni tsuite (Concerning the Promotion of the Teletopia Concept) (Internal Document of the Ministry of Posts and Telecommunications, unpublished).

Tisdell, Clem (1975) 'An Australian review of Japanese science and energy policy', in *Australia–Japan Economic Relations Project.* Canberra: Australian National University.

US International Trade Commission (1983) *Foreign Industrial Targeting and Its Effects on US Industries, Phase I: Japan.* Washington DC: US Government Printing Office.

Wakaki Shigetoshi (1986) *Nyu Baiotekunorojī Sangyō (New Biotechnology Industry).* Tokyo: Nihon Kogyo Shimbun.

Watanabe, Makoto (1985) 'Promotion of life science – What is Japan Doing?' in *Keidanren Review No. 95.* Tokyo: Nikkei Shinbun.

Yano, Toshihiko and Amadani, Naohiro (1984) *Haiotekunorojī No Sōzō to Katsuyō (The Creativity and Adoption of High Technology).* Tokyo: Daiichi Hooki.

Yoshida, Motoki (1984) *Reputo: Baiotekunorojī no Shinkaihatsu to Kadai (The Question of Biotechnology's New Development).* Tokyo: Japan Industrial Bank.

Zimmerman, Burke K. (1984) 'Trends in world biotechnology' in *Genetic Engineering and Biotechnology monitor.* Vienna: United Nations Industrial Development Organization.

Index

Abramovitz, Moses 57
advisory bodies (ABs) 15, 19–24, 63–6
agricultural chemistry 29–30
agriculture 108–9 nitrogen fixation 41; superplants 41–2
AIDS 32
Aji-no-moto 30
Akazawa Shoichi 75
Akio 43
Allen, G.C. 6
Allison, Graham 27
amino acids 30–1
Anderson, Alun 96
animals 43
antibiotics 30, 40
antibodies, monoclonal 31–2, 43
artificial insemination 43
Asahi Chemical 79, 97

Beggs, John 112
bioboom 34, 72–4, 125
bioindustry 33–4, 73
Bioindustry Advisory Council (BAC) 8, 23, 73–4
Bioindustry Development Centre (BIDEC) 34, 96–9, 117, 119
biomass 37, 43–4, 92
bioprocessing 29, 35
bioreactors 35
biosensors 40
biotechnology 1, definition 28–34; expectations 36–44; high tech society 46–61; international perceptions of 126–8;

opposition to 113–15; policy development 62–88; policy implementation 89–115; policy-making 5–27, 116–32; popular press 34–5; reasons for promotion 46–55; safety and regulations 44–5; science 35–6; state role 121–6
Biotechnology Research Advancement Institution (BRAIN) 105
Bosu no Sensei 65
Bosu Seiji 23–4
Boyer, Herbert 31, 97 Cohen-Boyer patent for rDNA technology 71
Brazil 44
bureaucracy 3, 18, dynamism 27, 89–90; *Nawabari Arasoi* 25–7; *Nebukai* 11–12; *Nemawashi* 18–19; politics 25–7; *see also* ministries; Triple N Synthesis

Calgene company 42
cancer 37–8, 50, 92
Cape, Ronald 7
cell-fusion technology 31, plants 42
Cetus corporation 7
Chakrabarty, Dr Ananda 71
cloning, antibodies 31–2; plants 41
Cohen, Stanley 31, 97
Cohen-Boyer patent for rDNA technology 71
competitiveness 3, 59–60
consensus 89, for high tech 53–4

151

Index

life sciences 66; RA projects
95; and rDNA 84–7; and
resources 109; Science Council
63, 66, 83; training in
biotechnology 107–8
Ministry of Health and Welfare
(MHW) 15, 82–3 Health and
Welfare Science Council 63;
Human Science Promotion
Foundation 95, 98–9; RA
projects 92, 95; and rDNA 88,
91; and resources 109
Ministry of International Trade
and Industry (MITI) 1, Agency
of Industry Science and
Technology (AIST) 75, 77, 79–
80; bioindustry 33–4, 73, 98;
Bioindustry Advisory Council
(BAC) 8, 23, 73–4;
biotechnology policy
development 70–81, 128, 130;
co-ordinating role 123–4;
expectations of biotechnology
36; Industrial Structure Council
20, 63 75; intermediaries 119–
20; next generation basic
technology project 74–81;
policy development 70–1;
policy-demand-incites-action
theory 71–4; RA projects 92,
95; and RAs 99–102; and
rDNA 88; and resources 109;
Shingikai 22–3; *Vision for the
1970s* 47–8; *Vision for the 1980s*
1, 6, 14, 46, 49, 53, 74–9
Ministry of Posts and
Telecommunications (MPT)
52–3 and New Key Tech Centre
103–5
Mitsubishi 79, 92, 97, 99, 104
Mitsui Toatsu 79
monoclonal antibodies 31–2, 43
monosodium glutamate (MSG) 30
Morishita Noboru 72
Moritani Masanoru 60
Murakami Yoichirō 106

Nakasone Yasuhiro, Prime
Minister 19, 50–1, 67
National Institute of
Agrobiological Resources
(NIAR) 108–9
National Research Institutes for
Joint Use by Universities
(JRIJUs) 108
Nawabari Arasoi – bureaucratic
politics 2–3, 25–7 supportive
role 128–9
Nebukai – cohesive conservative
coalition 2–3, 10–14, 117–18
bureaucracy-private sector
relations 12; and *Nemawashi*
119–21; political-bureaucratic
relations 11–12; political-
private sector relations 13;
united coalition 13–14
NEC 99
Nemawashi – consensus and
conflict 2–3, 14–24 advisory
bodies 19–24; bureaucracy 18–
19; and *Nebukai* 119–21
New Key Technology Promotion
Centre 103–5, 119–20
Next Generation Basic
Technology project (NGBT) 7,
74–81, 92, 101, 111, 120, 131
Nihon Digital Equipment
Corporation 104
Nihon Keizai Shimbun 93
Nippon Roche Company 104
Nippon Telephone & Telegraph
(NTT) 53, 104
nitrogen fixation 41

Ōbei 128
Ohira, Prime Minister 19, 87
Okimoto, Daniel 12, 93, 123
opposition to biotechnology 113–
15
organizations 16–17

P4 114–15
patents and RAs 101–2; rDNA 71
Pempel, T.J. 27
penicillin 30, 40

Index

state-led capitalism approach 2, 3–4
Stockwin, J.A.A. 6
subsidies and grants 91–2
Sumitomo Chemical 79, 97
Suntory 51, 102
superplants 41–2
Suzuki Eiji 80–1
Swanson, Robert 71

Takara Distillers 72
Takeda Chemical 104
Tanabe Pharmaceuticals 30, 102, 124
Tanaka Masami 97–8, 103
targeting of business areas 3, 5–6
tariffs 5, 90–1
tax incentives and borrowing 93
technology, and economic growth 55–60; imported 54–5; see also biotechnology; high tech
technopolis plan 52
teletopia 52–3
tissue plasminogen activators (tpa) 38
Toa Nenryō 104
Toray Industries 104
Toshiba 99
training 50, 106–7
Triple N Synthesis 3–4, 9–10, 116–32 effectiveness of policy 130–2; Nawabari Arasoi 2–3, 25–7, 128–9; Nebukai 2–3, 10–14, 117–18; Nebukai and Nemawashi 119–21; Nemawashi 2–3, 14–18; state and industry 129–30; state role 121–6
tropical diseases 39–40
Tsukuba Life Sciences Centre 109, 114–15

Uchida Hisao 45, 85–8
Unilever Company 41
United Kingdom (UK) research associations 99–100; Spinks Report 79; see also Europe
United States of America (USA), Cohen-Boyer rDNA patent 71–2; education and training 50, 106–7; International Trade Commission (USITC) 5–6; US-Japan relations 50–1; legal test cases 45; medical research 37; National Institute of Health (NIH) 44–5, 85, 86; new technology 54, Office of Technology Assessment (OTA) 6–7, 33, 131; research grants 91; San.Gaku.Kan 94; view of Japanese biotechnology 126–8; view of Japanese policy 5; and VLSI project 91, 99, 120
universities 50, 106–8 and ABs 23–4, 65; and industry 112–13; and state 105–12; see also San.Gaku.Kan
urokinase 38

vaccination 39–40, 92, animals 43
Very Large-Scale Integrated Circuit (VLSI) project 80, 91–2, 99, 120

Watanabe Itarō 65
Watson, J.D. 31, 65
Weber, Max 18
West, the 128 see also Europe; USA

Yoshida Motoki 92
Yoshida Shigeru 50
Young, Frank 88

For Product Safety Concerns and Information please contact our EU
representative GPSR@taylorandfrancis.com
Taylor & Francis Verlag GmbH, Kaufingerstraße 24, 80331 München, Germany

9 781032 897912